A IS FOR ARSENIC

A IS FOR ARSENIC

The Poisons of Agatha Christie

Kathryn Harkup

BLOOMSBURY
sigma

Bloomsbury Sigma
An imprint of Bloomsbury Publishing Plc

50 Bedford Square
London
WC1B 3DP
UK

1385 Broadway
New York
NY 10018
USA

www.bloomsbury.com

First published 2015

British Library Cataloguing-in-Publication Data
A catalogue record for this book is available from the British Library.

Library of Congress Cataloguing-in-Publication data has been applied for.

ISBN (hardback) 978-1-4729-1130-8
ISBN (trade paperback) 978-1-4729-1131-5
ISBN (ebook) 978-1-4729-1129-2

2 4 6 8 10 9 7 5 3 1

Typeset in Bembo Std by Deanta Global Publishing Services, Chennai, India

Molecular diagrams by Julia Percival
Illustrations by Neil Stevens
Printed and bound in Great Britain by CPI Group (UK) Ltd, Croydon CR0 4YY

Bloomsbury Sigma, Book Six

That was the beginning of the whole thing. I suddenly saw my way clear. And I determined to commit not one murder, but murder on a grand scale.

Agatha Christie, *And Then There Were None*

Contents

Contents

Dame Agatha's Deadly Dispensary

She hath pursued conclusions infinite
Of easy ways to die.

William Shakespeare, Anthony and Cleopatra

Dame Agatha Mary Clarissa Christie (1890–1976), the 'Queen of Crime', holds the Guinness World Record as most successful novelist of all time. She has been outsold only by the Bible and by Shakespeare (and is more widely translated than the Bard); Christie is also the author of the world's longest-running play, *The Mousetrap*, and created not one but two of the best-known fictional detectives, Hercule Poirot and Miss Marple. Praise, prizes and awards have been heaped on Christie for her work, and her books and plays are still loved by millions.

Many have attempted to divine the secret of her success. Christie always considered herself a 'popular' writer, and acknowledged that she did not produce great works of literature or deep insights into the human condition. Nor did she revel in gore, or try to shock her readers with gratuitous violence. Christie created many corpses in the pages of her books, but the reactions they produce are most likely to be curiosity and a smile at the prospect of clues, red herrings and brilliant deductions. She was a teller of tales, an entertainer, and a poser of seemingly insoluble puzzles.

Christie's detective stories repeatedly demonstrate that she was a master of misdirection. She presented clues fairly and in plain sight, often drawing attention to them, but safe in the knowledge that most readers would form their own, wrong, conclusions. When the murderer is finally revealed readers are generally left kicking themselves that they didn't spot the obvious, or they cry foul and go back to the beginning, only to discover the clues were there all along.

Christie took advantage of her detailed knowledge of dangerous drugs to help develop her plots. She used poisons in the majority of her books, far more than any of her contemporaries, and with a high degree of accuracy, but she did not expect the reader to have detailed medical expertise. The symptoms and availability of drugs are succinctly described in everyday language, and somebody with a degree in toxicology or medicine has no real advantage over any other reader.* An understanding of the science behind the poisons Christie used only gives a better appreciation of her cleverness and creativity in plotting.

A poisoner's apprentice

Agatha Christie's knowledge of poisons was certainly exceptional. Few other novelists can claim to have been read by pathologists as reference material in real poisoning cases (see page 267). Several of the people who kindly read chapters of this book in its early stages have asked me 'How did she know all of this?' The answer is that her knowledge came from direct experience with poisons and a lifelong interest in the subject, though not in the criminal sense.

In the First World War, Christie volunteered as a nurse at her local hospital in Torquay. She enjoyed the work but when a new dispensary opened at the hospital it was suggested that she might work there. Her new role required further training, and Christie also needed to pass examinations to qualify as an apothecary's assistant, or dispenser, which she did in 1917.

Then and for many years afterwards, doctor's prescriptions were made up by hand in a chemist's shop or hospital dispensary. Poisons and dangerous drugs were carefully weighed out and checked by colleagues before being dispensed. Innocuous

*Chemists, pharmacists and similarly qualified people may be able to eliminate certain possibilities early on in the novels, but the revelation of the murderer is just as surprising to them as it is to anyone else.

ingredients such as colouring or flavouring could then be added according to personal taste. As Christie explained in her autobiography, this resulted in many people returning to the pharmacy to complain that their medicine didn't look right, or didn't taste as it usually did. As long as the drug in question had been added in the correct dose all was well, but accidents sometimes occurred.

In order to prepare for the Apothecaries Hall examination Christie was tutored in practical, as well as theoretical, aspects of chemistry and pharmacy by her colleagues at the dispensary. In addition to her work and tutoring at the hospital, Agatha received private tuition from a commercial pharmacist in Torquay, a Mr P. As part of her instruction, one day Mr P. showed her the correct way to make suppositories, a tricky task that required some skill. He melted cocoa butter and added the drug, then demonstrated the precise moment to turn the suppositories out of the moulds, box them up and label them professionally as *one in one hundred*.* However, Christie was convinced that the pharmacist had made a mistake and added a dose of *one in ten* to the suppositories, ten times the required dose and potentially dangerous. She surreptitiously checked his calculations and confirmed the error. Unable to confront the pharmacist with his mistake, and frightened of the consequences of dispensing the dangerous medicine, she pretended to trip and sent the suppositories crashing to the floor, where she trod on them firmly. After she had apologised profusely and cleared up the mess, a fresh batch was made, but this time at the correct dilution.

Mr P. had carried out his calculations using the metric system, at a time when the imperial system of measurement was vastly more common in Britain. Agatha Christie didn't trust the metric system because, as she said, 'The great danger … is that if you go wrong you go ten times wrong.' By putting the decimal point in the wrong place, Mr P. had made a serious miscalculation. Most pharmacists at the time were much more

---

*i.e. one part drug per hundred in total.

familiar with the traditional apothecaries' system, which measured out drugs in units called grains.*

It wasn't just Mr P.'s inattention to detail that troubled Christie. One day, he pulled a brown lump from his pocket and asked her what she thought it might be. Christie was perplexed, but Mr P. explained that it was a lump of curare, a poison originally used by hunters in South America on the tips of their arrows. Curare is a compound that is completely safe to eat but deadly if introduced directly into the bloodstream. Mr P. explained that he carried it around with him because 'it makes me feel powerful'. Nearly fifty years later, Christie resurrected the deeply disconcerting Mr P. as the pharmacist in *The Pale Horse*.

*

By 1917, Christie had written some poems and short stories, a few of which had been published. And then, after reading *The Mystery of the Yellow Room* by Gaston Leroux, Christie thought she would try to write a detective novel herself, and said as much to her sister, Madge. But Madge, a more successful writer than Agatha at the time, stated that it would be very difficult, and bet her that she wouldn't be able to do it. It was not a formal bet, but nonetheless it spurred Christie to write. It was while working as a dispenser that she found she had the time to think about the plot and her characters, and, being surrounded by poison bottles, she decided that poison would be the means of murder.

The resulting novel was *The Mysterious Affair at Styles*, and Christie demonstrated her detailed knowledge of strychnine throughout the book. However, she had to wait a few years and try a number of publishing houses before the novel was finally accepted in 1920. After publication Christie received her most cherished compliment when it was reviewed in the *Pharmaceutical Journal and Pharmacist*. 'This novel has the rare

---

*Christie used grains throughout her stories, but in this book I'll give equivalent measurements in grams (g) or milligrams (mg, thousandths of grams); one grain is equivalent to 64.79891mg.

merit of being correctly written,' the reviewer stated. He believed the author must have had pharmaceutical training, or had called in an expert.

A criminal career

The publication of *The Mysterious Affair at Styles* was the start of a long and very successful career, but it was only after publishing three novels that Christie acknowledged that she might be a professional writer. She maintained her interest in poisons and drugs throughout her writing life, and only reluctantly used guns in her work – she freely admitted to knowing nothing about ballistics. The scientific details of her chosen poisons were well researched. She built up a considerable medico-legal library over the years, with the most well-thumbed book in her collection being *Martindale's Extra Pharmacopoeia*.

During the Second World War Christie volunteered again as a dispenser at University College Hospital, London. After renewing her training she put in regular hours at the dispensary, working two full days a week plus three half days and Saturday mornings. She also filled in when other workers were unable to get to the hospital. Her work at the hospital kept her up to date with new developments in drugs and pharmaceutical practice. At this time an increasing number of standard formulations were pre-prepared, and Christie found herself with plenty of time on her hands to invent new stories and work out fiendishly deceptive plots.[*]

Christie also corresponded with experts to check her facts. For example, in 1967 she wrote to a specialist asking about the impact of putting thalidomide in birthday-cake icing – how long would it take to make an impact? How many grains would be needed? However, this idea was never used in any of her stories.

---

[*]Christie was able to write twelve complete novels during the war years.

Christie was writing during the so-called Golden Age of Detective Fiction. In the 1920s and 30s detective fiction was a serious business. In 1928 Ronald Knox[*] (1888–1957) wrote his 'Detective Story Decalogue' – a series of ten rules that crime writers were expected to adhere to in the spirit of fairness to the reader. These were:

1. *The criminal must be someone mentioned in the early part of the story, but must not be anyone whose thoughts the reader has been allowed to follow.*
2. *All supernatural agencies are ruled out as a matter of course.*
3. *Not more than one secret room or passage is allowable.*
4. *No hitherto undiscovered poisons may be used, nor any appliance which will need long scientific explanation at the end.*
5. *No Chinamen must figure in the story.*
6. *No accident must ever help the detective, nor must he ever have an unaccountable intuition that proves to be right.*
7. *The detective must not himself commit the crime.*
8. *The detective must not light on any clues that are not instantly disclosed to the reader.*
9. *The stupid friend of the detective, the Watson, must not conceal any thoughts that pass through his mind; his intelligence must be slightly, but very slightly, below that of the average reader.*
10. *Twin brothers, and doubles generally, must not appear unless we have been duly prepared for them.*

Agatha Christie broke nearly all of the rules, most spectacularly in *The Murder of Roger Ackroyd*, which caused consternation at the time of publication, and columns of newspaper screed declaring Christie to be a cheat. Today the novel is regarded as one of the best detective stories of all time. When not actually breaking the rules, she strained them to their absolute limit. In spite of this Christie was one of the founder members of The Detection Club, a dining club for writers of detective fiction

[*]Knox was a priest, theologian, BBC broadcaster and writer of detective stories. His books featured the detective Miles Bredon.

whose members included G. K. Chesterton, Dorothy L. Sayers and the writer of the Decalogue himself, Ronald Knox. Knox's rules were adopted by the club's members as a kind of code of ethics among detective-fiction writers. Members also had to swear an oath as part of an elaborate initiation ceremony.

> *Do you promise that your detectives shall well and truly detect the crimes presented to them using those wits which it may please you to bestow upon them and not placing reliance on nor making use of Divine Revelation, Feminine Intuition, Mumbo Jumbo, Jiggery-Pokery, Coincidence, or Act of God?*

Christie showed only slightly more regard for the oath than for Knox's rules, yet she still managed to maintain fairness to her readers. Christie was proud of the fact that she never 'cheated'. The clues were presented, but it was for the reader to spot and interpret them correctly.

In terms of poisons, Christie invariably played with a straight bat. She never used untraceable poisons; she carefully checked the symptoms of overdoses, and was as accurate as to the availability and detection of these compounds as she could be. But there were a few notable exceptions. Serenite (*A Caribbean Mystery*), Benvo (*Passenger to Frankfurt*) and Calmo (*The Mirror Crack'd from Side to Side**) are drugs that are pure Christie inventions, though the properties she attributed to them are very similar to those of barbiturate drugs. In fairness to Christie, she only used one of her invented drugs to kill a character, in *The Mirror Crack'd from Side to Side*; otherwise, these drugs were not critical to the plot.

Her use of poison was not just a convenient way to dispose of a character. Although her novels are liberally sprinkled with classic poisons such as arsenic and cyanide, Christie used a huge variety of killer compounds in her novels – too many to fit into this book. Many of the poisons she described were the drugs she was familiar with from her dispensing days. Toxic

---

*Shortened to *The Mirror Crack'd* for publication in the United States.

compounds and chemicals such as strychnine, phosphorus, coniine and thallium were still in use in 1917 in medical preparations. They have long since disappeared from the *British Pharmacopoeia*, due to their high toxicity and low therapeutic value. Other compounds, however, such as morphine, eserine, digitalis, atropine and barbiturates, still have applications in modern medicine. As the physician and founder of toxicology Paracelsus (1493–1541) pointed out, 'Poison is in everything, and no thing is without poison. The dosage makes it either a poison or a remedy.' Christie understood this well, and made use of unusual and unexpected poisons, such as nicotine and ricin, to great effect. The symptoms, availability and detection of the poisons contributed clues and plot points to her stories. For example, the brilliantly plotted novel *Five Little Pigs* makes use of hemlock – the way it acts on the body, its taste, and the time it takes to act all match perfectly with the timeline of the novel (see pages 136–140).[*]

Real-life inspiration
Agatha Christie did not just rely on an accurate and detailed knowledge of poisons. She read about real crime extensively, and was well versed in the sensational murders of the past. She referred to many real-life murders and poisoners in her books, killers such as Herbert Rowse Armstrong, Frederick Seddon and Adelaide Bartlett. She even used the circumstances of murder cases as inspiration for her plots.

Mrs McGinty's Dead is a novel based around the infamous murderer Dr Hawley Harvey Crippen. He was found guilty of poisoning his wife, and was hanged in 1910. Human remains were discovered buried in the cellar of the Crippens' London home. The portions of flesh, wrapped in Crippen's pyjamas, were found to contain lethal quantities of hyoscine hydrobromide. Meanwhile, Crippen had fled aboard a ship

---

[*]Incidentally, for much of the time that Christie was developing the plot she was planning to have the murder victim shot, though how this would have worked in the final novel is difficult to imagine.

bound for Canada with his mistress, Ethel le Neve, who was disguised as a boy. The disguise didn't fool the ship's captain, who alerted the British police by wireless telegram. Inspector Dew boarded a faster ship and was able to arrest Crippen and le Neve when their ship docked at Montrose. In *Mrs McGinty's Dead* several murders are committed to hide the murderer's secret – their mother was the mistress of a man who had killed his wife and buried her in a cellar.

The novel *Ordeal by Innocence* tells the story of Jacko Argyle, who was found guilty of murdering his mother. Years later, after Jacko had died in prison, a stranger turned up at the Argyle house with proof that Jacko was innocent. If Jacko did not kill his mother, who in the family did? The story was inspired by the Bravo case, a real-life poisoning that occurred in 1875. Charles Bravo married a wealthy young widow, Florence Ricardo, after a whirlwind courtship. Just four months into their married life Charles was taken ill after eating dinner with his wife and her live-in companion, Jane Cox. He died three days later after being attended by Dr James Gully, his wife's former lover. Post-mortem analysis revealed he had been poisoned by a single dose of antimony. An inquest into the death decided on an open verdict, though it was widely suspected that Charles Bravo had committed suicide.

Subsequent reports in the press revealed that Jane Cox and Charles Bravo had been on bad terms, and that Cox had overheard an argument between the married couple over Florence's association with Dr Gully. A second inquest was opened which effectively became a trial of the two women. A verdict of 'wilful murder' was returned, but with insufficient evidence to suggest who had administered the fatal dose of antimony. By this time the two women were no longer friends. The common suspicion was that Mrs Bravo had laced her husband's wine with poison and attempted to cast suspicion on her companion. Charles Bravo's murderer was never identified. To quote Agatha Christie, 'And so Florence Bravo, abandoned by her family, died alone of drink, and Mrs Cox, ostracised, and with three little boys, lived to be an old woman with most of

the people she knew believing her to be a murderess, and
Dr. Gully was ruined professionally and socially.' As Christie
eloquently put it, 'Someone was guilty – and got away with it.
But the others were innocent – and didn't get away with
anything.'

*

I first read Agatha Christie's books when I was a teenager. I
loved the stories, but I doubt if I appreciated the scientific
content at the time. Re-reading the novels and short stories
during my research for this book has only increased my
appreciation, not only of Christie's scientific knowledge, but
also of the way she incorporated it into her work. Many find
science off-putting, but Christie explained all the detail
necessary to understand the significance of a poison without
distracting the reader from the plot. In this book, I'll examine
fourteen of the poisons Christie employed during her writing
career, as well as the real-life cases that could have inspired her,
or may have been inspired by her work. It is a celebration of
Christie's inventiveness, her brilliant plotting, and her attention
to scientific accuracy.

ARSENIC

Murder is Easy

The poison of Kings and the King of poisons.

Anon.

The name 'arsenic' has become almost synonymous with poison – it could be argued that it represents the gold-standard of criminal poisoning. Arsenic has a long and illustrious history of murder and assassination, stretching from the time of the Ancient Greeks to the present day. This is often the poison people most associate with Agatha Christie, but in fact only eight characters, in four novels and four short stories, were dispatched using this infamous element and some of these die 'off stage' with little description of their symptoms. They form a relatively small proportion of the more than 300 characters Dame Agatha bumped off in her career. In fact, her use of arsenic is relatively low-key given its infamy. However, it does get a mention in many of her books, and some of these often fleeting references reveal her deep knowledge of the poison.

The 1939 novel *Murder is Easy** features an arsenic murder with some details of symptoms, as well as discussions about how the arsenic could have been administered. The novel has a stereotypical 'Agatha Christie' setting, with mass murder being carried out in a quiet English village. Luke Fitzwilliam, a retired detective, takes on the task of solving the crimes. Fitzwilliam is drawn into the case by an elderly spinster, Lavinia Pinkerton, whom he meets on a train journey to London. She tells him she is on her way to Scotland Yard to report three suspicious deaths in her home village: Amy Gibbs, who died after drinking hat paint (yes, a paint for changing the colour of a hat) that she mistook for cough medicine; Tommy Pierce, who died after falling from a roof when he was cleaning windows; and Harry Carter, who fell from a bridge and drowned after a night out drinking. Did they fall, or were they pushed? Miss Pinkerton is convinced that these were no accidents, and she assures Fitzwilliam that Dr Humbleby will be the next victim.

Initially, Fitzwilliam dismisses the old lady's story, but when he later reads both Miss Pinkerton's and Dr Humbleby's obituaries in the newspaper he decides to take a closer look. He travels to Miss Pinkerton's village and investigates every recent death in the village, of which there have been a number. All of them appear to have been accidents, or due to natural causes, but the unprecedented demands on the local undertaker have meant that it was either a very unlucky village or something more sinister was going on. One of the deaths Fitzwilliam is most concerned about is that of Mrs Horton, the wife of Major Horton, who died the previous year after a long illness. She had been in hospital for some time suffering from acute gastritis. Although her symptoms could be explained by natural causes, they could also have been due to arsenic poisoning …

*Entitled *Easy to Kill* in the United States.

The arsenic story

Arsenic (As) is the fourteenth most common element in the Earth's crust, though it occurs naturally as a compound rather than as the pure element. It was first isolated in the thirteenth century, and was found to be a grey metalloid.[*] The name 'arsenic' comes from the Persian word *zarnikh*, which means 'yellow orpiment', a brightly coloured compound of arsenic and sulfur.[†] *Zarnikh* was then translated into the Greek word *arsenikon*, which was related to another Greek word, *arsenikos*, meaning 'masculine' or 'potent', before we finally arrive at arsenic. When people refer to arsenic as a poison they are usually referring to 'white arsenic' or arsenic trioxide (As_2O_3), or other deadly compounds of arsenic. In its pure elemental form, arsenic is far less toxic than arsenic trioxide, because the body cannot easily absorb it.[‡]

The poisonous properties of arsenic compounds have been known since at least the time of Cleopatra. When the Egyptian queen decided to end her life she wanted to ensure her death would cause her as little pain as possible, and that she would leave an attractive corpse. It is said that she tested various poisons on her slaves, and watched the results. One of the poisons she tested was arsenic, but it was clearly too unpleasant a way to die, so she opted for the asp (though this would have been far from pain-free, and her cadaver would also have needed some cosmetic retouching).

[*]On the boundary between a metal and a non-metal, with characteristics of both.

[†]Christie of course used the British spelling, 'sulphur', whenever she mentioned the element or its compounds in her work. Since 1990 the scientific community has agreed that the spelling to use is, in fact, 'sulfur', which is the one I'll be using in this book; this will look perfectly acceptable to American eyes but may jar a little with British ones, and I can only apologise. The *Oxford English Dictionary* still lists the British spelling as 'sulphur', so there's a good case for using either.

[‡]For the rest of this chapter, the word 'arsenic' refers to arsenic trioxide unless otherwise stated.

In more recent times, arsenic-poisoning was a popular murder method in Renaissance Europe, and was a particular favourite of the Borgia family. It was claimed that the Borgias spread arsenic on the entrails of a slaughtered pig, which were then left to rot. The resulting mess was gently dried to a powder which they called *La Cantarella*, a pale solid that was added to food or drink. If the arsenic did not claim the victim, the toxins from the rotting entrails would probably finish them off. The benefits of using arsenic were twofold. First, arsenic had no taste that could alert the potential victim to their being poisoned. And second, the symptoms of arsenic poisoning are very similar to those of food poisoning, cholera and dysentery, all of which have been common at various times through the ages.

Throughout the sixteenth and seventeenth centuries, poisoning was considered a peculiarly Italian art, owing in part to the reputation of the Borgias,[*] as well as Toffana (a professional poisoner who labelled her deadly cosmetics with the images of saints) and the Council of Ten, one of the governing bodies of Venice. The Council of Ten maintained its position of power by killing off potential rivals, and even went to the lengths of actively advertising for poisoners, as well as maintaining its own reliable stock of poisons for nefarious purposes.

By the seventeenth century the popularity of arsenic poisoning had spread to the French royal court. Members of the aristocracy were found to be conspiring with La Voisin, a notorious poisoner who was also alleged to have taken part in black masses. The investigation became so widespread, and involved so many prominent figures in French society, that a special court was convened, the *Chambre Ardent* or 'Burning Court', named after the method commonly used for executing those who the twelve judges found guilty. To save embarrassment and potential repercussions the court met in secret and reported only to the King. 'Human life is up for sale, and cheaply,' wrote Nicolas Gabriel de la Reynie, one of the judges; 'Poison is the sole solution to most family problems.'

[*]Although they were in fact Spanish in origin.

And arsenic, of course, was the poison of choice; its use had become so common that it was referred to as *poudre de succession* – 'inheritance powder'.

Prior to the seventeenth century, many powerful and influential people took to employing official tasters, and there was understandable caution over who was allowed to prepare their food and drink. There are many stories of methods of murder that circumvented the tasters. Tales of gloves and riding boots laced with poison that would kill by contact with the skin are probably exaggerated, but tests have shown that a poisoned shirt is at least a theoretically viable method of administering arsenic. The garment in question would have the tail soaked in a solution of arsenic before being allowed to dry. The material would appear slightly stiff but otherwise there would be no obvious sign of anything wrong with the fabric. Contact via the skin of a naked bottom could potentially allow enough arsenic to be taken in to kill, especially if a blistering agent was added to the arsenical mixture to break the skin and allow faster absorption into the bloodstream.

Arsenic poisoning was long the preserve of the rich and powerful. Those on more restricted means had to find other ways to kill each other. However, the Industrial Revolution brought with it huge demands for metals such as iron and lead, and when extracted from the ground as ores these metals are often contaminated with arsenic. To obtain the pure metal, the ore was roasted in fires, and the arsenic reacted with oxygen in the air to form arsenic trioxide. This would condense in the chimneys as a white solid, which had to be periodically scraped off to prevent the chimney from getting blocked. Instead of dumping the white arsenic as waste, industrialists realised a profit could be made by selling the arsenic as poison, for rats, bedbugs, cockroaches or any other vermin infesting the home (including humans). Prices plummeted, and soon anyone and everyone could afford enough arsenic to dispatch an unwanted relative or inconvenient enemy.

Unsurprisingly, the number of arsenic poisonings began to rise. Anyone reading a nineteenth-century British newspaper would think that arsenic murders had reached epidemic proportions, with working–class women being the most likely culprits. In reality the number of poison trials was very low; even at the peak of the media hysteria there were only two or three trials per year in the whole of England and Wales. There was sometimes suspicion when there were sudden deaths, but this was often fuelled by local gossip from folks who might have had reasons to hold a grudge, and by sensationalist reports in the newspapers. For example, of the 20,000 suspicious deaths in England in 1849, 415 were linked to poison, but only eleven of those were possibly murder, and not all resulted in a guilty verdict. What did not help these early cases was that, if arsenic was involved, the victim's symptoms could be attributed to natural causes, and there was no way to detect arsenic in the body. It became apparent that something needed to be done, and various methods were trialled to allow the detection of arsenic in human tissue. None of these early methods was reliable, and they did not produce results that could easily be discussed, or even shown in a court case. The case of John Bodle illustrates this well.

In 1832 James Marsh (1794–1846), a British chemist, was asked to investigate the death of an 80–year–old farmer named George Bodle. Marsh found arsenic in the dead man's intestines and in a cup of coffee he had drunk from, but the samples he prepared for the trial did not keep too well, and the jury found Marsh's technical descriptions of his experiments incomprehensible. As a consequence the suspect, John Bodle – the farmer's grandson – walked free. John Bodle later confessed to the murder, but he could not be re-tried. Marsh was furious and set about devising a test for arsenic that even the most stupid jury member could comprehend. He wanted jury members to see the arsenic for themselves.

Marsh made a U-shaped glass tube with one end open and a tapered nozzle at the other. In the nozzle he suspended a piece of zinc. The fluid to be examined was placed in the open end, and acid was added. When the fluid level reached the zinc even a minute trace of arsenic would be converted to arsine

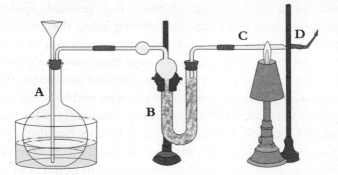

Marsh Test apparatus as it was in 1921. (A) flask containing zinc and acid for generating hydrogen; (B) calcium chloride for drying the gas produced; (C) glass tube; (D) arsenic mirror. From An Introduction to Chemical Pharmacology *by Hugh McGuigan.*

gas (AsH_3) that would be ignited as it left the nozzle. A cold porcelain bowl was held against the flame; pure metallic arsenic was then deposited on its surface. The apparatus would be further refined, but it was still in use in Agatha Christie's time – indeed, in preparation for her apothecary's examinations she and a colleague practised the Marsh test using their 'Cona' coffee machine, blowing it up in the process.

The Marsh test was first used in a criminal trial in 1840, by the celebrated toxicologist Mathieu Orfila (1787–1853). Orfila had been asked to investigate the death of Monsieur Charles Pouch-Lafarge. When Marie Capelle married Charles in 1839, both believed they were marrying into serious wealth. In reality Marie had a modest dowry, but she had ideas above her station. She had been educated at elite schools and believed herself to be descended from royalty. Charles claimed to be a wealthy iron-founder, but he really lived in a tiny hamlet, in a damp, rat-infested house on a dilapidated estate, part of which he had converted into a foundry; the expense of the conversion had left him almost broke. The marriage did not get off to the best of starts, and things did not appear to improve until Marie persuaded her husband to change his will in her favour. Charles went away to Paris over the Christmas period to seek out financial backers for a new business venture, and Marie sent him a parcel as a Christmas gift. Members of the Lafarge household had witnessed Marie placing five small cakes in a box, with a portrait of herself and a loving letter. When the package arrived in Paris it contained one large cake that made Lafarge very unwell when he ate it. He recovered enough to travel back home, but was again taken ill and died shortly afterwards. Arsenic poisoning was suspected, and Marie was known to have purchased arsenic, supposedly to kill the rats in the house. Orfila was called in to determine if arsenic had indeed been the cause of Charles's death. Orfila's testimony and the results of his Marsh test provided enough evidence for the jury to find Marie guilty of murder.*

There are still doubts over Marie's guilt in the Lafarge case. No one could prove that she had switched the cakes, or even that she had had the opportunity to do so. Another toxicologist, François-Vincent Raspail (1794–1878), also threw doubt on the forensic evidence. Raspail showed that the zinc Orfila had used when carrying out the Marsh test had been contaminated with arsenic and would have given a positive test for the poison even if *none* had been present in Charles's remains. Raspail's evidence came too late, though, and Marie had already been sentenced to life imprisonment before his arrival at the court. Raspail had highlighted the only real drawback of the Marsh test – it was perhaps a little too sensitive. Being able to detect 0.02mg of arsenic would normally be considered an advantage in forensic science, but this element is widespread in the world, and was particularly so in nineteenth-century

---

*This case may have inspired Christie's use of arsenic-laced cake in *After the Funeral*. In the novel a slice of arsenic-laced wedding cake was delivered to a suspect in a murder case. The suspect ate some of the cake but placed the remainder under her pillow, following the tradition that she would then dream of her future husband. Although the suspect grew very ill she survived because she didn't eat all of the cake.

European households. Arsenic was soon found to be almost everywhere.

As its name suggests, white arsenic is a white powder. It is very similar in appearance to sugar or coarse flour, and mistakes sometimes occurred. Food adultery was fairly common in Victorian England. Sweet-makers would add 'daft', an inert substance such as plaster of Paris or powdered chalk, to bulk up their sweets, because it was cheaper than sugar. In 1858 one sweet-maker in Bradford went to what he thought was a barrel of daft, scooped up the white powder within, and proceeded to make his sweets. Unfortunately, he had taken the powder from a barrel containing white arsenic. When some children ate the sweets the mistake was soon realised and the deadly candy was recalled, but not before 200 people had become seriously ill and twenty had died. Extraordinarily to modern eyes, no one was prosecuted over the incident.

By contrast, in 1836 Eliza Fenning, a cook, was executed for the attempted poisoning of the household she worked in. The whole family, and Eliza herself, had become seriously ill after eating dumplings she had prepared (though they all recovered). A packet of arsenic had disappeared from the house some weeks earlier. Eliza was convicted on the flimsiest of evidence. Others in the household had had the opportunity to add poison to the dumplings, or – perhaps more likely – it was all a terrible mistake.

Eighteenth-century industrialists found other uses for the growing amount of waste arsenic from the iron smelters, besides rat poison. Several arsenic compounds are brightly coloured, and these have been used as pigments for thousands of years, such as orpiment (As_2S_3), an intense yellow, and realgar (AsS), a ruby red mineral. In 1775 Scheele's green ($CuHAsO_3$), invented by Carl Wilhelm Scheele (1742–1786), was added to the list of arsenic-based pigments. These arsenic compounds were a huge improvement on the vegetable dyes that had been used in the past as their colours did not fade so readily and

they were cheap and easy to manufacture. The great popularity
of the colours red and green in Victorian England meant that
arsenic was used to dye almost anything and everything, from
wallpapers and clothes to toys and even food, such as sweets
and cake icing.*

Arsenic dyes in wallpaper were directly dangerous only to
those manufacturing the paper, who would be exposed to the
arsenic dust. In the home it was clear that bedrooms with
arsenical wallpaper had fewer bedbugs. This was initially seen
as a bonus, and sales increased. The problem was that whatever
was affecting the bedbugs soon began to affect the human
occupants of the room, too. To stick wallpaper to a wall a
simple flour paste was used. In the damp climate of the British
Isles this provided the perfect environment for mould to grow.
Mould is also adversely affected by arsenic but some moulds
could adapt to their environment by chemically processing and
removing it. In 1893 Bartolomeo Gosio (1863–1944) was the
first to show that *Penicillium brevicaule* (now known as
Scopulariopsis brevicaulis) was attacking the starch paste and
releasing an arsenic gas, which he could not identify but which
had a distinctive garlicky smell. The gas became known as
Gosio gas; it was in fact trimethylarsine gas ($As(CH_3)_3$),
identified in 1933. Trimethylarsine is highly toxic, and
recommendations were made to reduce the amount of arsenic
used in wallpapers. Unfortunately these recommendations
came too late for one famous Frenchman.

There has been a lot of speculation about the death of
Napoleon Bonaparte in 1821. During the last months of his
exile on St Helena he was very unwell, and was attended by
many doctors, both French and British. The emperor was
suffering from severe stomach pains, and medical treatment

*Agatha Christie makes reference to Scheele's green and its use in
wallpaper in the novel *They Came to Baghdad*. When one of the
characters is taken ill with 'Bad gastro-enteritis', arsenic poisoning is
suspected. 'I'm wondering', says Sir Rupert, 'if it might be a case of
Scheele's green …'

seemed to have little effect on him. When he died, seven doctors attended the autopsy and concluded that Napoleon had died of stomach cancer, but rumours of poisoning were swift to circulate. As you might expect, the French accused the British and the British accused the French. Very little could be done to confirm or deny poisoning at the time as reliable tests were not available.

In the 1960s samples of Napoleon's hair, cut from his head shortly after death as mementoes, were analysed for arsenic content. Unusually high levels of arsenic were discovered, opening up questions as to how it might have got there. One theory was that it came from his wallpaper; when a sample of the wallpaper from his bedroom was discovered in the 1980s, analysis showed significant levels of arsenic, $0.12g/m^2$.[*] In 1893 a detailed study showed that wallpapers containing between 0.6 and $0.015g/m^2$ arsenic could cause health problems, and even values as low as $0.006g/m^2$ were potentially hazardous.[†] St Helena had a warm, damp climate likely to encourage the growth of mould in wallpapers, but even so this is unlikely to have generated enough trimethylarsine to kill Napoleon. The wallpaper certainly may have contributed to his poor health, though, and he did what anyone else would do when they feel ill, he called in a doctor. Unfortunately the doctors who attended Napoleon did little to help, and they introduced more toxic compounds into his body in the form of medicines, though probably not actually with the intention of poisoning him.

The range of medicines available to a nineteenth-century physician was limited, and those available were generally used because they had been shown to produce an effect on the human body, such as purging (vomiting and diarrhoea) or sweating. Any recovery from an illness was usually incidental, and clinical trials or follow-up consultations were practically unheard of. Sick people often got better in spite of their

[*] g/m^2 = grams per square metre.
[†] At the time the recommended safe limit was between 0.001 and $0.005gm^{-2}$.

doctor's ministrations, rather than because of them. The compounds that produce the most dramatic effects on the human body are often highly toxic, and before the twentieth century a doctor's medical bag would almost certainly contain compounds that are considered highly dangerous today.

One medication in common use in the nineteenth century was Fowler's solution, a tonic prescribed for the treatment of a variety of illnesses. Fowler's solution was introduced into the *British Pharmacopoeia* in 1809, initially as a treatment for malaria. It had no taste, which made it preferable to the bitter quinine medicine usually prescribed. The number of prescriptions for Fowler's solution increased, along with the number of complaints it was prescribed for, from skin conditions to asthma. The key ingredient in Fowler's solution was, of course, arsenic, in the form of potassium arsenite (K_3AsO_3).[*]

Finding arsenic in a Victorian corpse would, therefore, not be that surprising. The prosecution in a criminal poisoning case had to prove not only that arsenic was the cause of death, rather than incidental, but also how the arsenic had been obtained and administered to the victim.

The Arsenic Act was passed in 1851 in an attempt to regulate and control the sale of arsenic.[†] The act made it a legal requirement for sales to be recorded in a register along with the name of the purchaser, the quantity bought and the purpose it was to be used for. The Act also required any arsenic not used for medical or agricultural purposes to be 'coloured' with either soot or indigo dye, to reduce the risk of mistakes such as the sweet incident. Unfortunately, there were large loopholes in the Act; for example, there were no restrictions on who

---

[*]Charles Darwin is known to have taken Fowler's solution, initially as an undergraduate to treat his eczema, but he continued dosing himself for much of his adult life. This may go some way to explaining the poor health he experienced as an adult.

[†]There doesn't seem to have been an equivalent law passed in the United States. A survey in 1877 revealed that the sale of poison was still unregulated in the US, though things are rather different today.

could sell arsenic compounds, and anyone determined to murder another human being was unlikely to have any qualms about recording false information in the poison register. Over time the rules on the sale of arsenic, and other poisons, were tightened up. The sale of poisons was restricted to a few professions and shops, such as pharmacies, and anyone buying a poison had to be known to the pharmacist, or vouched for by someone who knew both the pharmacist and the purchaser. Even if criminal poisonings were carried out there should in theory have been a system for tracking the poisoner through the poison registers. It was still frighteningly easy to get hold of arsenic, though, and a huge number of 'legitimate' uses could be recorded in the poison register, but it was up to the prosecution to prove that the accused's intentions had been otherwise.

A case could be complicated further if the accused used the 'Styrian defence'. This was a legal argument used to explain the presence of high levels of arsenic in a corpse. In 1851, a report appeared in a Viennese medical journal about men from the Austrian region of Styria who regularly ate arsenic. They would crunch lumps of arsenic trioxide between their teeth, or grate it onto their toast two or three times a week. They would start with a lump the size of a grain of rice, and gradually increase the dose until they could eat quantities normally considered lethal with apparent impunity. The reason for this strange choice of dietary supplement was because they said it gave them 'wind', by which they meant that they could breathe more easily while doing hard physical labour in the thin mountain air. The arsenic also gave the men more physical bulk and clearer skin, making them more attractive. Women in the region used arsenic too, as it gave them a more curvaceous figure and a 'peaches and cream' complexion.

The arsenic was indeed killing off any bacteria that might have caused spots and blemishes, but it was also triggering oedema – retention of fluid in the muscles – and vasodilation of the capillaries under the skin to give the rosy-red cheeks. The habit might be expected to have made the arsenic-eaters feel ill, but some complained they actually felt unwell when they missed a dose. On first appearances it might look as if they were developing a tolerance of arsenic – which would have been handy for anyone who suspected a relative of trying to bump them off. However, these individuals were not developing a true tolerance. Eating large quantities of arsenic was possible because it was swallowed in relatively large lumps, rather than as a fine powder or dissolved in a liquid. Much of the arsenic would have been excreted before it could be absorbed into the bloodstream.

After reading about the Styrian arsenic-eaters, their attractive appearance and apparently excellent health, some people across Europe and America also started taking arsenic. Arsenic would be used as a beauty treatment, applied directly to the skin, or dissolved in water and drunk in small quantities to improve general health. Agatha Christie had read about arsenic-eaters, and described a widow in *Evil Under the Sun* who had the advantage, or disadvantage, of having an arsenic-eater for a husband, which enabled her to walk free from her trial for his murder.

Arsenic is a cumulative poison, and levels would have gradually increased inside an arsenic-eater's body until dangerous or potentially lethal levels were achieved. Even if their death was not attributed to arsenic poisoning, arsenic-eaters were relatively easy to identify post-mortem as the arsenic acted as a preservative in the body, killing the bacteria that would normally drive the process of decomposition. The burial tradition in Styria involved the removal of a corpse from a grave after twelve years; land for graves was in short supply, so the bones of the deceased were removed to a crypt, and the plot was left vacant for the next occupant. Arsenic-eaters were often found so well preserved, even after twelve years, that they were recognisable to family and friends when they were disinterred. The presence of arsenic in corpses may lie behind some vampire legends, which began in central and eastern Europe.

The preservative properties of arsenic led to its use in the embalming process, until it was realised that this would mask any arsenic that was present owing to a homicidal poisoning. Arsenic was banned from use in embalming, to be replaced by formaldehyde. Even then, the problem of arsenic contamination of a corpse did not go away. Arsenic is a common mineral in soil, and a dead body could potentially absorb it from the ground it was buried in.

Arsenic bonds very strongly to sulfur atoms, of which there are many in the body, particularly in the hair. This provides a useful record of arsenic exposure over the lifetime of the hair, as arsenic is deposited at the roots within hours of ingestion. As the hair grows, the arsenic is retained at a fixed position. Hair grows at a fairly regular rate, approximately one centimetre (0.4 inches) per month, so a timeline of exposure can be built up by sequential analysis of strands of hair. It also means that the hair of a corpse lying in fluid containing arsenic will soak it up like a sponge, and store it; this results in higher concentrations in the hair than in the fluid. During post-mortem examinations care had to be taken that hair on the corpse was not allowed to come into contact with fluid from the body in case it artificially raised the arsenic level in the hair, giving the impression of long-term exposure. In cases of exhumation the same care had to be taken when removing the body from the ground, and samples of soil from around the burial site had to be collected and analysed.

In the Victorian era, cases of arsenic poisoning became difficult to prove, and the potential complexity of the situation is illustrated by the Maybrick case. In 1889, 50-year-old James Maybrick fell ill with stomach pains and violent vomiting. His 26-year-old American wife, Florence, nursed him devotedly. The couple had recently made up after a violent falling-out over a liaison Florence had had with a friend of her husband. James had also had a string of affairs, but it was Florence's infidelity that resulted in her getting a black eye, and he cut her out of his will. James was something of a hypochondriac, and took a lot of patent medicines to treat himself. During his last illness James asked his wife for his powders, and Florence added some to a bottle of meat juice for her husband to drink, as was his usual custom.

This time James got no better. His family soon arrived to see that he was receiving proper medical attention. Florence was not popular in the household after a letter written to her lover was intercepted, in which she had written that her husband was 'sick unto death', and she was effectively banished from the sick room. When James subsequently died a couple of weeks later, suspicion immediately fell on Florence.

At Florence's trial, the prosecution was able to show that she had bought arsenical fly-papers, which she claimed she had intended to use to prepare a tonic for her skin; she had run out of her usual facial wash and had decided to try and make her own. When Florence purchased the fly-papers she also bought a lotion containing benzoin and elderflower water – the usual ingredients for the skin lotion. Soaking a fly-paper in cold water extracted three-quarters of a grain of arsenic (not enough to kill), but boiling water extracted almost all of the arsenic (over two grains, close to a lethal dose), and it also extracted the colouring in the paper. The fly-papers were sold in packets of six and were clearly labelled as poisonous. The quantity of arsenic in each fly-paper varied, but every one of a packet that was analysed for the trial was found to contain enough arsenic for at least one lethal dose.[*]

In fact, Florence Maybrick had no real need to buy fly-papers. There was already plenty of arsenic in the Maybrick household. The police had discovered bottle after bottle of cosmetics and patent medicines in their search of the house, with many containing arsenic. There was enough of the poison to kill 50 people, but there was one place where there was relatively little – inside James Maybrick's body. Medical men were called by Florence's defence to testify that James Maybrick had died of natural causes. No one had witnessed Florence administering arsenic to her husband, and for several days before his death she had no contact with her husband, his food

---

[*]Agatha Christie was well aware of the Maybrick case and she describes soaking fly-papers to obtain poison in her novel *The Mysterious Affair at Styles*.

or his medicines. The jury, however, thought there was enough evidence to convict her and found Florence guilty. Her death sentence was commuted to life imprisonment because doubts remained over whether James had died of arsenical poisoning, rather than whether it was Florence who had poisoned him. Florence protested her innocence throughout the 14 years she was imprisoned, and after her release appears to have led a blameless life.

When Agatha Christie began her writing career, arsenic compounds could still be obtained relatively easily in the form of medicinal 'tonics', pesticides and weedkillers, but arsenic was slowly phased out of use over the first half of the twentieth century as alternatives were found for rat poisons and herbicides. Some specialist industrial uses remain, though, but there is only one remaining medical use for arsenic trioxide, for the treatment of acute promyelocytic leukemia. However, this treatment is not without risk from arsenical poisoning.

How arsenic kills

The toxicity of arsenic trioxide and related compounds stems from their disruption of basic chemical processes within the body. Arsenic compounds are readily absorbed through the skin, lungs and gastrointestinal tract, and most poisoners have exploited this by adding arsenic to food, drink or medicines.

Arsenic compounds occur in two forms, arsenates and arsenites, and they interact with the body in different ways. Arsenates (AsO_4^{3-}) are structurally and chemically similar to phosphates (PO_4^{3-}) and the body is unable to distinguish between the two. Phosphates perform many vital roles in biology, from strengthening bones to forming the backbone of DNA's double helix. These chemical units are also involved in vital chemical processes within cells, one of the most important of which is the transfer and storage of energy. In the home, electricity is the form of energy we use to power our gadgets and appliances. Inside the body, energy from the food we eat and oxygen we breathe is used to produce a chemical called adenosine triphosphate, or ATP. The transfer of phosphate units from ATP to other molecules makes them chemically more reactive, and allows reactions to occur under the relatively mild conditions experienced in most biological systems. Arsenate compounds kill because they can substitute for phosphate in ATP. Arsenate is less chemically reactive than phosphate, and consequently the chemical reactions it is involved with are slowed and may even stop – and this is very bad news.

Arsenic trioxide, arsine gas and the arsenic pigments such as Scheele's green are all *arsenite* compounds rather than arsenates; they kill because of a different chemical interaction with the body. To describe how arsenite compounds do this I'll use the example of arsenic trioxide, as this has been the arsenite compound most commonly employed in murder over the centuries.

The first symptoms of arsenic poisoning – severe vomiting and abdominal pain – appear approximately thirty minutes after ingestion, and are triggered by the irritant effects on the tissues of the stomach. If the victim is lucky much of the poison will be eliminated from the body at this point, but the unlucky ones will have absorbed a fatal amount, approximately 100–150mg, into the bloodstream. Many victims of arsenic poisoning in the past survived for weeks, probably because they purged much of the poison in vomit and diarrhoea. Many nineteenth-century arsenic poisoners were very attentive nurses, ensuring that they were on hand to administer extra doses to achieve their desired result.

The inflammation of tissues in the stomach and intestine caused by arsenic trioxide may be visible to pathologists at post-mortem examination, but it is not the cause of death. The violent vomiting and profuse diarrhoea, however, can cause dehydration, and this may kill if the fluids cannot be replaced. However, it is arsenic trioxide's disruption of biochemical processes in the body that usually proves terminal for the victim.

Chemical reactions inside the body are carried out by proteins called enzymes. These are large molecules made from strings of amino acids that wrap and twist into precise shapes that allow them to carry out their function. Enzymes are able to carry out specific chemical reactions, on compounds generically referred to as substrates, at a very fast rate. One analogy often used to describe their operation is the 'lock and key' theory. The enzyme is the lock, the substrate is the key and the active site where reactions occur is the space where the key fits into the lock. Very few keys will open more than one lock because of the close match needed between the two complementary elements. The enzyme alters the substrate or 'key' in a chemical reaction, changing its size and shape so it no longer fits in the 'lock'. The substrate then dissociates from the enzyme, leaving it free for the next substrate to bind to.

Some of the amino acids that form proteins and enzymes contain sulfur atoms, and these atoms often form crucial chemical bonds that hold an enzyme in shape.[*] Arsenic (in the form of arsenite) bonds very strongly to sulfur atoms, and this can distort the shape of the enzyme or 'lock' and thereby stop it working. Once arsenic compounds have entered the bloodstream they can be distributed around the body and potentially affect any sulfur-containing enzymes or proteins that they encounter.

Because of the huge number of enzymes and their diverse roles in the body, arsenic poisoning can present many different symptoms. The amount of arsenic administered has a dramatic effect on the symptoms displayed, and the ultimate cause of death. Massive doses of arsenic, ten times the minimum lethal dose, will produce symptoms of violent gastroenteritis, vomiting and intense stomach pain, along with copious amounts of watery or bloody diarrhoea. Later the skin becomes cold and clammy, blood pressure drops and death comes from circulatory failure within hours. Convulsions and coma may be seen, and these signal that the individual is very close to death.

[*]The sulfur is often in the form of a sulfydryl group, a sulfur atom bonded to a hydrogen atom (–SH).

Some of the many enzymes that arsenite compounds can disrupt are those involved in the energy process within cells. Without a supply of energy a cell cannot function, and it rapidly dies. When large numbers of cells die this leads to organ failure. Some cells have higher energy demands than others; for example, heart and nerve cells require more energy than red blood cells do. Other enzyme-regulated metabolic processes within cells are also susceptible to arsenic interference; there are many ways arsenic can cause cell death. If the poisoned individual survives for a day or two they grow jaundiced, and urine output reduces or stops because of damage to the liver and kidneys, the organs normally involved in detoxifying and eliminating poisons from the body.

Arsenic doesn't only kill by acute poisoning, as described above. It is also lethal following the slow accumulation of small amounts administered over a long period of time; this is known as chronic poisoning. At these lower doses, arsenic causes nausea and vomiting but also headaches, dizziness, cramp and variable paralysis that may progress over a period of several weeks. In addition, heart arrhythmias may occur. Death in chronic cases is due to multi-system organ failure; in such cases there can also be damage to the nerve cells of the central nervous system, specifically to the axons, the long parts of a motor neuron (a nerve cell that controls movement) that stretches from the spinal cord to the extremities, carrying messages. Symptoms that result from this damage include numbness or burning sensations in the hands and feet. Chronic poisoning may also damage the liver, kidneys and circulatory system. Regular doses of arsenic accumulate in areas of the body with high sulfur content, such as in the protein keratin, which makes up hair and nails and is also present in the skin. Over time the skin of an arsenic-eater, for example, would lose its 'peaches and cream' quality and become darkened (a process called hyperpigmentation), with horny or scaly patches on the palms of the hands and feet. Characteristic transverse white lines across the nails, known as Mees lines, would also appear. There is likely to be weight loss and, if the person survives all of this, arsenic can go on to cause cancer of the skin, lungs and liver.*

The increased risk of cancer from exposure to arsenic has been known about for more than a century. Sir Jonathan Hutchinson (1828–1913), a physician and an expert in dermatology, amongst other things, noticed an unusually high number of skin cancers in patients who had been prescribed arsenical medication for various illnesses. It is thought that arsenic disrupts the body's ability to repair DNA damage, though several mechanisms may be involved.

The consequences of long-term exposure to arsenic are amply demonstrated in places such as the Ganges Delta, where wells built by well-intentioned aid agencies were bored through arsenic-containing rocks. This led to mass low-level poisoning. The wells were built to prevent the spread of water-borne diseases such as cholera that were common in areas where surface waters were poorly maintained. The wells have saved lives through lowering incidents of cholera, but tens of millions are now at risk of arsenic exposure.

The body does excrete arsenic, but slowly. As long as arsenic is not ingested at a higher rate than it can be excreted all is well, and this is the case for most of us, most of the time. Arsenic is present in our environment, in the soil and in water, and thereby gets into our food supply, but it's generally in very small quantities that our bodies can manage.

The half-life (time taken for half the amount to disappear) of arsenic trioxide in humans is approximately ten hours. Arsenic trioxide is either excreted unchanged in urine or metabolised into other arsenic compounds. Methyl groups ($-CH_3$) are sequentially added to the arsenic molecule; it was once thought that this process detoxified the arsenic. In fact

---

*TV and films often portray poison victims swallowing a mouthful of tainted food, choking a bit and collapsing in seconds. This is far from accurate, and has often resulted in my shouting at the screen with frustration. I am surprised I haven't been banned from my local cinema yet.

many of the methylated arsenic compounds are just as toxic as arsenic trioxide, if not more so. The methylated compounds may result in a garlicky odour to the breath, similar to the odour of trimethylarsine produced by the mould growing on Victorian wallpaper. Methylated arsenic compounds have a half-life of around thirty hours. Therefore, excretion of 50 per cent of ingested arsenic can take between one and three days.

Is there an antidote?

A treatment for acute arsenic poisoning was first demonstrated in 1813 at the French Academy of Sciences, when chemist Michel Bertrand (1774–1857) ingested 5g of arsenic (around 40 times the lethal dose) along with some charcoal. He survived and showed none of the usual symptoms of arsenic poisoning, proving that the charcoal had somehow inactivated the arsenic. In fact the arsenic became trapped in tiny cavities within the charcoal, preventing it from being absorbed into the body. Studies have shown that charcoal is effective in absorbing many other poisons and it is still used as the first line of treatment in suspected poisoning cases, though it is only really effective if it is used relatively soon after ingestion.* Charcoal can be processed further to make it 'activated charcoal' by treating it with steam, carbon dioxide, oxygen, zinc chloride and sulfuric acid at high temperatures (260–480°C); this increases the number of pores (and therefore it can absorb more poison).

Once arsenic has been absorbed into the body charcoal cannot remove it, and alternative methods must be used. These were developed in response to the invention of Lewisite gas, an arsenic-based poison, in the First World War. British Anti-Lewisite (BAL), or dimercaprol, is known as a 'chelating' compound; a chelate wraps around a metal ion, such as arsenic, binding to it at several points to form a tightly bound metal-chelate complex. Once BAL has scavenged the arsenic from

---

*As recently as the early 1960s, employees at poison-treatment centres were known to start work by burning toast to prepare the day's supply of charcoal.

the body, the resulting arsenic-chelate can be excreted. Other chelating agents have been developed since then that can extract arsenic more efficiently; these are more specific for toxic heavy metals, and so have fewer side effects. Unfortunately, chelating agents aren't effective for *chronic* arsenic poisoning. The most effective treatment in these circumstances is simply to reduce exposure.

Some real-life cases

Agatha Christie referred to many real-life arsenic-poisoners in her novels, and used poisoning cases as inspiration for her plots. One poisoner she mentioned by name was Frederick Seddon, a particularly avaricious landlord who was found guilty of killing one of his tenants. Miss Eliza Mary Barrow, a wealthy spinster, died after Seddon had persuaded her to make over all her money to him. Suspicion was only aroused by Barrow's relatives, who were surprised to hear that not only had Eliza died, but that she had already been buried. Seddon even haggled over the price of the funeral. When the relatives enquired about the money they knew Eliza possessed, they were told by Seddon that there had been very little. In court the prosecution was able to show that arsenic was present in Barrow's body. Seddon claimed that Barrow must have got up and drunk from the dishes of arsenical fly-papers that had been placed in her room.

Another possible arsenic murderer, Madeleine Smith, also gets a mention in several Agatha Christie stories. In 1855, aged 20, the Glasgow socialite embarked on an affair with Pierre Emile L'Angelier. Smith had promised to marry L'Angelier, but her parents, unaware of her attachment, had found another eligible bachelor, William Harper Minnoch, and arranged a formal engagement with their daughter. Madeleine attempted to end the relationship with L'Angelier and asked him to return her love letters. He instead threatened to send them to her parents so her engagement to Minnoch would be broken and she would be forced to marry L'Angelier. Soon afterwards L'Angelier wrote in his diary that he felt unwell, often after seeing Madeleine. It was claimed she gave him cocoa to drink. He told his friends that he thought Madeleine was poisoning him, and around this time Madeleine was seen buying arsenic in a pharmacy. By March 1857 L'Angelier's illness had grown so severe that a doctor was called, and morphine was administered to alleviate his pain. This was to no avail – by the following morning he was dead. An autopsy revealed an enormous amount of arsenic in his stomach, more than 87 grains (approximately 5g). At the time no other case of murder had seen such large quantities present in the body (although suicide cases had). Such a large amount might be expected to be difficult to administer without the victim noticing but, as was discussed in the trial, up to 6g (20 to 60 times the lethal dose) of fine arsenic powder, mixed with two teaspoons of cocoa, plus milk or boiling water in a teacup, cannot be detected by appearance or smell. However, on cooling, the arsenic sedimented out of the cocoa, and curdled the milk.

Though Madeleine had a motive to kill her lover and was known to have purchased arsenic, the case was far from clear-cut. Madeleine claimed that the arsenic was for her complexion but, in accordance with the law, the arsenic she had purchased from the pharmacist was coloured with indigo dye. Madeleine must have known of a way of removing the dye from the arsenic before applying it to her skin (washing the arsenic in cold water would have done the trick). The arsenic found in L'Angelier's stomach was white arsenic with no trace of dye. The defence barrister prevented L'Angelier's diary from being entered as evidence in the trial, and the prosecution had failed to keep Madeleine's 200 undated love letters in their original envelopes. It became impossible to prove conclusively when the two had met, or that Madeleine had had the opportunity to administer the poison.

The scandal of pre-marital sex and murder ensured huge publicity for the trial, and even after the case against Smith was found 'not proven', speculation continued. On one side many believed that Madeleine had murdered her lover, but the prosecution was simply unable to prove when. Others believed that L'Angelier had committed suicide. After the trial Smith went to live in England under an assumed name and married an artist, George Wardle. Two children were born but after many years of marriage the couple separated. Madeleine moved to New York where she changed her name once more, and finally died in 1928.

Agatha and arsenic

The title of Agatha Christie's 1939 novel, *Murder is Easy*, is appropriate. It charts seven murders in a tiny English village in the course of just over a year. The methods used were varied, and chosen to look like accidents or natural diseases. The first victim, Mrs Horton, seemed to have died of acute gastritis after a long illness. Her death was all the more tragic as she had seemed to be getting better before a sudden and dramatic relapse. Even the doctor attending her had been surprised at the suddenness of her death but at the time there was no suspicion of foul play. Only a year later, when the village churchyard is filling up a little too rapidly, does anyone take a closer look at the circumstances of Mrs Horton's illness and death.

'Gastroenteritis' describes a set of symptoms – vomiting, diarrhoea and stomach pain – rather than a specific illness, and these are due to inflammation of the gut. The symptoms can be brought about by a large number of causes, usually a virus such as the norovirus or a bacterial infection, or, more rarely, by a parasite, but they can also be due to food intolerances – or even arsenic poisoning. An infection might normally be expected to clear up after a few days or a few weeks. Mrs Horton was described as having a long illness, so we can assume she was ill for at least a number of weeks and must have been suffering from chronic arsenic poisoning, with a larger dose administered shortly before she died.

Symptoms of chronic arsenic poisoning, Mees lines and skin effects such as pigmentation and dermatitis, for example, may not have had time to present themselves during Mrs Horton's illness. Nails grow at approximately 3mm per month, and though arsenic can be deposited in hair and nails within hours of ingestion, it would take several weeks for the deposit to grow out from the nail matrix and past the cuticle, where it would become visible. In the Agatha Christie novel *They Do It with Mirrors*[*] the would-be murderer carefully cuts the victim's nails so they cannot be analysed for arsenic. However, this plan would not have succeeded; the poisoning would have to have been carried out a long time in the past, or the whole nail would have had to be removed to prevent detection. Even then, the hair of the victim could also have been analysed and would have shown signs of arsenic.

Even with no obvious outward signs of arsenic poisoning Mrs Horton was convinced she was being poisoned by one of her nurses, and had her dismissed. No one else seemed to take her seriously, and dismissing the nurse did not improve her condition. Someone else was administering the arsenic.

There are plenty of suspects to consider, chief among them being Mrs Horton's husband, Major Horton. Agatha Christie may have taken her inspiration for this character from a real-life arsenic-poisoner, Major Herbert Rowse Armstrong, dubbed 'The Dandelion Killer'. In 1921 Armstrong was working as a solicitor in Hay-on-Wye. His wife had died after a long illness that involved her being treated in a lunatic asylum, shortly after she had made over all her property to her husband. Her treatment in the hospital appeared to have been successful and she was well enough to return home, where she suffered a relapse and died a month later. No suspicion was attached to her death, even though the doctor could not determine the exact cause. But Armstrong's behaviour after her death did cause some alarm.

Oswald Martin was a solicitor at a rival legal firm in Hay-on-Wye. Martin and Armstrong were both involved in a land dispute, and were acting for opposing sides. The legal arguments had been going on for some time and were becoming increasingly acrimonious, which made Armstrong's behaviour all the more surprising.

[*]Entitled *Murder with Mirrors* in the United States.

Martin received an invitation to tea at Armstrong's house, which he accepted. He ate a tasty buttered scone, which Armstrong had selected and placed on a plate. Later Martin became very ill, but he recovered. He then recalled receiving a box of chocolates, sent anonymously, which had made a guest of his very ill. Martin was by now suspicious, and he decided to alert the police. The chocolates were analysed along with a sample of Martin's urine, and both were found to contain arsenic.*

The results of the tests were enough to interest the police and they agreed to investigate, but did not want to alert Armstrong to what was going on. Tea invitations continued to pour in to the Martin household from Armstrong, and Martin did his best to find excuses. Eventually Armstrong was arrested for the attempted murder of his rival; after his wife's body was exhumed and his house searched for arsenic, he was tried for the murder of his wife. Arsenic was found in the body, and packets of arsenic were discovered in the house. Armstrong claimed the arsenic was for poisoning dandelions in his lawn. The jury didn't believe him, and he was hanged.

In *Murder is Easy*, Major Horton seems more concerned about the health of his dog than the death of his wife, but the lack of an appropriate emotional response is not evidence of murder. There are also plenty of other suspects to consider. Mrs Horton seems to have been popular; she had several visitors when she was ill, and one of them, Lydia Pinkerton, would enquire about the food and drinks she was being given. She was clearly suspicious, and the murderer later pushes Miss Pinkerton in front of a car before she can reach Scotland Yard to tell the police of her concerns, although luckily she has

---

*The Marsh test is equally effective on samples, such as urine, from living people; today, the test for arsenic would involve using atomic absorption spectroscopy, where a sample is placed in a flame that causes it to emit light in colours characteristic of the elements present.

already shared her suspicions with a retired detective, Luke Fitzwilliam.

Another resident in the village, Lord Whitfield, showed his concern for the invalid Mrs Horton and sent her peaches and grapes from his hothouse. Mrs Horton complained that the fruit was bitter but the nurse never repeated the assertion. A bitter taste is sometimes a sign of a plant-based poison such as morphine or strychnine, but the symptoms Mrs Horton suffered from were not consistent with this type of poisoning. Arsenic poisoning would explain the symptoms, and it has no taste. The fruit could have been naturally bitter and still have had arsenic added to it. Arsenic applied to the skins of the fruit would have shown as a white powder, which would probably have been washed off if it was noticed. Alternatively, a solution of arsenic could have been injected into the flesh of the fruit. Relatively small doses could have been administered this way over a number of weeks.

Another method that could have been employed was adding arsenic to the patent medicines Mrs Horton was known to take. These medicines were supplied to Mrs Horton by the local antiques dealer, who appeared to have no motive for killing Mrs Horton, but his dabbling in black magic certainly made him a suspicious character. By the 1950s arsenic was no longer being used in tonics as its effects were 'unpredictable and uncontrollable' but in 1939 these remedies were still available, though their use was declining. Getting hold of arsenic in 1930s England would have been considerably more difficult than 50 or even 20 years earlier, but not impossible. No mention is made in the book of how the poison was obtained, but in 1935 ten grains of weed-killer contained seven grains (454mg) of arsenic, enough to kill two or three people. The weedkiller would have had a bright blue dye added to it to prevent accidental ingestion, and the dye would have been clearly visible if weedkiller had been added to Mrs Horton's food, drinks or patent medicines.

In *Murder is Easy*, Luke Fitzwilliam, the retired detective, can only speculate over the cause of Mrs Horton's death. He is not investigating any of the crimes in an official capacity and therefore cannot order an exhumation and post-mortem, which could have confirmed his suspicions. Even a year after burial, arsenic would have been easy to detect in a corpse. Hair takes a long time to decompose, and it would have held on to the arsenic deposits from chronic poisoning for many more years. Fortunately the murderer confesses, and even goes on to explain how the deed was done; the poison was added to Mrs Horton's tea by one of her visitors. Arsenic trioxide is poorly soluble in cold water, but it is much more soluble in hot water. By dissolving the arsenic in tea the killer was able to ensure that no suspicious gritty powder was left at the bottom of the cup.

The Labours of Hercules

Belladonna, n. In Italian a beautiful lady; in English a deadly poison. A striking example of the essential identity of the two tongues.

Ambrose Bierce, *The Devil's Dictionary*

Belladonna is a poisonous plant with a long history of use by humans as a beauty aid, as a medicine and as a murder weapon. All three applications were described by Agatha Christie in several of her stories, resulting in two attempted murders and one that was successful. In some respects belladonna is the perfect poison, as the plant grows in the wild. The principal poisonous component of belladonna is a chemical called atropine, which leaves no signs at post-mortem and is widely distributed throughout the body. Atropine is also rapidly broken down after death and may no longer be present at all a few weeks after burial, making it difficult to trace.

However, the bitter taste of this poison alerts many would-be victims, and the symptoms of atropine poisoning are easily recognised and treated.

Belladonna takes a starring role in *The Cretan Bull*, the seventh of the twelve short stories that make up the 1947 collection *The Labours of Hercules*. The premise of the book is that the great Belgian detective, Hercule Poirot, has decided to retire from detective work, but before he moves to the country to grow 'vegetable marrows', he elects to take on twelve cases that have a connection to the labours of Hercules, his namesake in classical Greek mythology. The legendary Cretan Bull was seduced by Pasiphae in disguise, leading to the birth of the Minotaur (in some versions of the tale the bull was Zeus in disguise, and he did the seducing, with Europa the target of his affections). The seventh labour of Hercules (the Greek, not the Belgian) was to overpower the Cretan Bull, which was playing havoc with crops and boundary walls on Crete, and bring it to King Eurytheus.

In Agatha Christie's version of the myth, Hercule Poirot captures Hugh Chandler and returns him to his fiancée, Diana Maberly. Hugh is a 'young bull of a man' whose erratic and frightening behaviour certainly causes havoc at his home, Lyde Manor. Hugh's symptoms of vivid dreams and hallucinations are blamed on a family history of insanity. His madness has almost pushed him to the point of suicide before Poirot's intervention. There is plenty of insanity on display, but it perhaps isn't Hugh who is afflicted.

The belladonna story
Atropa belladonna, more commonly known as deadly nightshade, is one of the most poisonous plants growing wild in Britain. Both the scientific and common names highlight the plant's deadly characteristics. The name *Atropa* is from Atropos, the third sister of the three Fates in Greek mythology; Clotho spun the thread of human life, Lachesis measured the thread, and Atropos cut it. *Atropa* is also the origin of the name of belladonna's most toxic component – atropine. 'Belladonna' is

Italian for 'beautiful lady', and the plant is so named because the berries of the plant were used by Renaissance women to make themselves more attractive. The juice would be squeezed from the berries and applied directly to the eye using a feather. Atropine in the berry juice would cause the pupil to dilate.[*] Belladonna extract could dilate the pupils for up to three days but could also cause blurred vision, and extensive use could result in blindness. A more beneficial modern use of belladonna extracts is in ophthalmic applications, since dilating the pupils allows a better examination of the interior of the eye.

Agatha Christie made use of belladonna's effect on the pupils in a few of her stories, as a means of disguise. The idea was that very wide pupils would make the eyes appear darker, though it would have no effect on the colour of the iris. In *The Big Four* Hercule Poirot uses belladonna in his eyes and sacrifices his moustache to disguise himself as his fictional brother. This somewhat implausible trick seems to fool the villains. A similar ploy is used in *Three Act Tragedy*.[†]

Belladonna is a member of the family Solanaceae, which includes other notorious species such as mandrake and datura. All these plants have their place in witchcraft, medicine and myth. Other members of the family are considerably less terrifying – potatoes and tomatoes, for example. Even so, when tomatoes were first introduced into Britain people recognised their similarity to deadly nightshade and refused to eat them, thinking the fruits were poisonous. Displays of tomato-eating were arranged to reassure the public.

Perhaps the most famous of the poisonous Solanaceae is the mandrake. This plant is mentioned in the Bible and in several Shakespeare plays, and was even grown in the greenhouses at Harry Potter's school, Hogwarts. It was once believed that the mandrake was the living link between plants and animals

---

[*]The same effect is achieved by candlelight, because the pupil opens to allow more light into the eye to compensate for low light levels, hence the romance of a candle-lit dinner.

[†]Entitled *Murder in Three Acts* in the United States.

because underneath the green foliage its bifurcated root looks like a pair of legs. Close examination of the roots may tell you whether a particular plant is a mandrake or a ladydrake, and some roots were cut and carved to emphasize the distinction. Mandrakes contain several toxic compounds, which are present in all parts of the plant except the fruit. In the past the roots were sold for medical uses such as easing childbirth, inducing sleep and numbing pain before surgery. Compounds within the mandrake would have been effective in these applications, though by using only crude extracts of the plant, there would have been little control over potency or side effects. Mandrake roots were also sold as charms to increase fertility, and as an aphrodisiac, which would not have been effective at all. Mandrake harvesters ran a considerable risk in obtaining these roots. Legend has it that the plant gives a deadly shriek when it is uprooted, so painful and piercing it could kill a man. Anyone wishing to uproot the plant was advised to block their ears and tie a piece of rope from the plant to a starving dog. The dog would then be lured away with some appetising morsel, pulling up the mandrake behind it.

Agatha Christie mentions belladonna in several of her novels, but not mandrakes. She seemed to favour another member of the Solanaceae, datura, to inspire particularly villainous poison plots. Datura's poisonous compounds are found primarily in the flowers and seeds. The plant has a variety of common names, including thorn apple, because of the appearance of the fruit, and moonflower, because it flowers at night. One species of datura, *Datura strammonium*, known as jimsonweed ('James-Town weed'), was responsible for a mass poisoning of soldiers in Jamestown, Virginia.

The James-Town Weed (which resembles the Thorny Apple of Peru, and I take to be the plant so call'd) is supposed to be one of the greatest coolers in the world. This being an early plant, was gather'd very young for a boil'd salad, by some of the soldiers sent thither to quell the rebellion of Bacon (1676); and some of them ate plenti-fully of it, the effect of which was a very pleasant comedy, for they

*turned natural fools upon it for several days: one would blow up a
feather in the air; another would dart straws at it with much fury;
and another, stark naked, was sitting up in a corner like a monkey,
grinning and making mows [grimaces] at them; a fourth would
fondly kiss and paw his companions, and sneer in their faces with a
countenance more antic than any in a Dutch droll.*

*In this frantic condition they were confined, lest they should, in
their folly, destroy themselves – though it was observed that all their
actions were full of innocence and good nature. Indeed, they were
not very cleanly; for they would have wallowed in their own excre-
ments if they had not been prevented. A thousand such simple
tricks they played, and after eleven days returned [to] themselves
again, not remembering anything that had passed.*

(from Robert Beverly's *The History and Present State of
Virginia*, 1705)

In Haiti datura is known as the zombie cucumber, because of
its use in zombie powders. Making a Haitian zombie is a two-
stage process. Stage one uses a powder whose principal
component is pufferfish. The poison found in pufferfish,
tetrodotoxin, relaxes muscles by blocking the action of the
nerves that stimulate them. The muscles for breathing can
become paralysed, making the victim appear dead, even though
their heart may continue to beat very faintly and they are fully
conscious. There is no antidote to tetrodotoxin, but poison
victims can recover if their breathing is supported artificially.
Stage two of the zombie-making process uses a second powder,
administered after the victim revives from the effects of the
first. The principal component of the second powder is datura.
The hallucinogenic properties of compounds in datura can
make an individual susceptible to suggestions, and thereby
controllable. By careful dosing and maintenance of a controlled
diet an individual can be sustained indefinitely in a stumbling,
stupefied state.

The word 'datura' is Hindi in origin. In India the plant is
associated with both poisons and aphrodisiacs. It is used in the
Indian traditional medicine system, Ayurveda, and in rituals
and prayers to Lord Shiva. Datura has also been used in India

for less beneficial applications, including deliberate self-poisoning. Exact numbers are difficult to obtain but they run into the thousands (more than 2,700 in a 15-year period). Fortunately, the majority of these attempts are unsuccessful, but around one in ten poisoning cases results in death. It was the stories of datura's use in India that inspired Agatha Christie and her fictional poisoners in *A Caribbean Mystery* and *The Labours of Hercules*. There are several compounds within datura and other deadly Solanaceae that can affect the human body, but the dominant toxic compound in belladonna is atropine.

Atropine

Atropine is classed as a plant alkaloid. These compounds, when dissolved in water, generally produce alkaline solutions. They also tend to have a bitter taste. Many plant alkaloids have significant effects on the human body, and have found their way into medicine. Atropine is described as a tropane alkaloid because it contains the tropane chemical group (see Appendix 2). Tropane-based compounds are found in a wide range of plants, and their effects on the human body differ hugely.

Members of the Solanaceae family contain atropine and another closely related tropane compound, scopolamine, also known as hyoscine. Scopolamine was allegedly used by Dr Hawley Crippen to murder his wife Cora (see page 16), and by Agatha Christie to dispose of Sir Claude in her first play *Black Coffee*. The chemical differences between scopolamine and atropine are minor, and their effects on the body tend to be very similar.

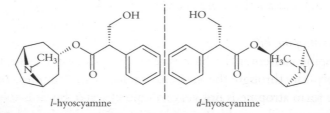

l-hyoscyamine   d-hyoscyamine

The two chiral forms of hyoscamine, which together make up atropine.

Atropine was first isolated from belladonna in 1831 by Heinrich F. G. Mein (1799–1864) and this plant, along with jimsonweed, remains the source of atropine for medical use today. Atropine is actually a mixture of two different forms of a chemical called hyoscyamine. These are called *l*-hyoscyamine and *d*-hyoscyamine; the two forms are mirror images of each other. Compounds such as these are described as 'chiral', and they are like a left hand and a right hand. Hands have identical components (fingers, thumb, palms and so on) but they are arranged slightly differently on each hand, forming mirror images that cannot be superimposed onto each other (hence the labelling in chiral compounds: *l*- for *laevo*, 'left' in Latin, and *d*- for *dextro*, 'right'). The two forms are chemically identical and have the same physical properties, such as melting point and solubility in water, but they differ in their interactions with other chiral molecules. Think of the difference between putting a left-handed glove on your left hand and putting the same glove on your right hand.

Biology is very good at producing just one hand of a chiral compound rather than the other, and our bodies are full of 'handed' molecules. The effects of drug molecules are caused by their interaction with other chemicals in the body; consequently, left- and right-handed forms of drug compounds can have very different effects in the body. In the case of *l*- and *d*-hyoscyamine most of the biological activity at low doses is thought to come from the *l*- version. At lethal doses, both forms appear to have equal potency. Individual sensitivity to atropine varies widely; some victims have died as a result of a dose of 10mg while others have survived 1,000mg, but it is generally agreed that toxic symptoms start to appear at doses of 5–10mg and a lethal dose is around 100mg.

How atropine kills

Atropine can enter the bloodstream by injection, ingestion or absorption through the skin and mucous membranes. In its pure form atropine is not very soluble in water, but it is soluble in fats and oils, so this form is more easily absorbed through the skin. Atropine in medication, in the form of eyedrops or for injections, is normally administered as a salt to improve its solubility in water.* Converting atropine into a salt does not alter its effect as a drug, just the ease with which it can be absorbed. For medical applications atropine is normally converted to its sulfate salt, which is readily absorbed through the gastrointestinal tract and mucous membranes but which tends not to pass through the skin.

Once absorbed into the bloodstream, atropine is quickly distributed through the body. It interacts with one of the two branches of the autonomic nervous system.† One of its branches, the sympathetic nervous system, is responsible for the body's 'fight or flight' responses to perceived danger; the other, the parasympathetic nervous system or PN, enables the body to 'rest and digest', and regulates the production of fluids such as tears, saliva and bronchial mucus. The PN does this by sending a chemical messenger (or neurotransmitter) called acetylcholine from its nerve cells to the target organs.

Acetylcholine is released from nerve endings that dock at specific receptors on adjoining organs or nerves. They work a little like a child's shape-selector toy – a neurotransmitter such as acetylcholine fits into a receptor and triggers a response in the target organ or nerve. In our analogy, pushing the correct shape into the hole would result in a light coming on. Other chemicals or 'shapes' can also fit into the receptors. A chemical that binds and activates a receptor is known as an *agonist*. However, some shapes may get jammed in the hole, so that the light won't come on and the hole is now blocked. A molecule that binds but does not produce a response is known as an *antagonist*.

The different types of neurotransmitter receptor are often named after compounds that interact with them. Muscarinic receptors, which are found throughout the body, can be

---

*Adding an acid to the alkali atropine will make the corresponding salt. For example, by adding hydrochloric acid, atropine chloride is formed.
†This is responsible for controlling automatic functions outside conscious control, such as the heart rate, salivation and certain reflex actions, such as coughing.

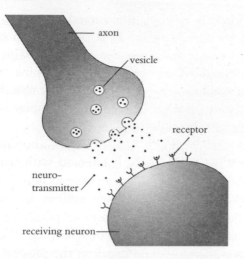

How neurotransmitters work. Signals are transmitted across the gap between nerves (or nerves and muscles) using chemicals called neurotransmitters. They are released from the nerve-ending of one and dock at receptors on the other, triggering a response.

activated by muscarine, a toxic substance found in several kinds of mushroom. Molecules of muscarine will compete with acetylcholine to bind to and activate these receptors. Atropine also competes with acetylcholine to bind to muscarinic receptors. However, it *doesn't* activate the receptor and is therefore an *antagonist*. By blocking the effects of acetylcholine, atropine hinders the actions of the parasympathetic nervous system (the PN, the 'rest and digest' mechanism).

Atropine – in small doses – can be used to dry up the secretions normally produced as a result of PN activity. In the past it was used to treat the symptoms of hay fever and colds by stopping the production of mucus, and atropine is still sometimes used today in cough syrup, where it has the added effect of relaxing the muscles used to cough. Atropine is also injected before operations to dry up bronchiole secretions in the lungs, which might otherwise block airways during surgery,* and it is sometimes administered as a palliative to stop

*Although *l*-hyoscyamine, one of the 'hands' of atropine, is increasingly used, probably because it has fewer side effects.

the 'death rattle' in dying patients that can be so distressing for the family; the 'death rattle' is caused by an accumulation of saliva in the throat and upper chest, which the patient is no longer able to clear for themselves.

The PN stimulates digestion by increasing the production of gastric juices, and by activating muscles that move food through the gut. It also controls the excretion of waste from the bowel and bladder. Atropine is therefore sometimes prescribed in the treatment of irritable bowel syndrome, and in the past it was given to children to stop bed-wetting because it blocks signals to the muscles controlling the bladder.

Even when atropine is given in therapeutic doses (5–10mg) there can be worrying side effects in sensitive individuals, such as dilated pupils, blurring of vision and an increased heart rate. But these side effects can be utilised, in the treatment of eye conditions such as miosis,* for example. Another condition treated with atropine is anterior uveitis, which causes inflammation of the iris. Atropine paralyses the muscles controlling the iris, causing the pupil to dilate and thereby relieving the pressure. Blurred vision will occur initially because the muscles that control the lens inside the eye are also temporarily paralysed, but these effects soon wear off. However, the effect on pupil-size can last for days. The dose is low, and application is localised (with drops administered directly to the eye) to minimise side effects, though some individuals have reported suffering hallucinations as a result of using these eyedrops.

✽

The feelings of disorientation and even hallucinations from atropine use are due to its effects on the other main part of the nervous system, the central nervous system, or CNS. Hallucinations from the use of atropine are visual and realistic – the appearance of faces, trees and snakes are often

---

*Miosis is an excessive constriction of the pupil that can be caused by several medical conditions.

reported, as opposed to the psychedelic images and patterns often experienced by those under the influence of drugs such as LSD. People often describe their visions in terms of looking at the world through a sheer fabric or piece of tissue paper, which possibly reflects a combination of the effects of the drug on the lens in the eye as well as on the brain. In medical conditions, patients given a dose of atropine often appear as if they are daydreaming – it is difficult to get and maintain their attention. Initially patients are docile, but this may progress to paranoia as they realise that what they think they are seeing isn't really there; there are difficulties with the perception of time, and a sense of disorientation is common. The hallucinogenic effects of atropine can last for up to twelve hours.

Atropine has a short half-life in the body of around two hours. Most of the compound is excreted unchanged in urine, but a significant proportion is metabolised by enzymes in the liver. Even so, it takes a long time for all the atropine to disappear from the body, and the effects of the drug can persist for days. This also means that regular small doses can accumulate in the body, leading to chronic poisoning. At high doses atropine causes hot, dry, red skin (sometimes a rash appears, usually on the upper half of the body), dry mouth, a rapid pulse and breathing, urinary retention, muscular stiffness, fever, convulsions and coma, in addition to the disorientation, hallucinations and delirium seen at smaller doses. The most worrying effects are on the heart and on breathing.

The effect of atropine on the heart is caused by the presence of muscarinic receptors in that organ. The sympathetic nervous system acts to increase heart rate, while the PN slows it. By blocking muscarinic receptors in the heart, atropine serves to diminish the signals from the PN telling it to slow down; this means that atropine can be used as an antidote for overdoses of drugs that slow the heart (it will also help raise the blood pressure and pulse). However, if too much atropine is administered, atropine poisoning can result. Fortunately it is fairly easy to diagnose atropine poisoning, by following this mnemonic: 'hot as a hare, blind as a bat, dry as a bone, red as a beetroot and mad as a hatter'. A patient who survives for longer than 24 hours will probably recover.

Is there an antidote?

When low doses of atropine have been administered, simply removing the source of the poisoning is often enough for the victim to make a full recovery from the effects, though it may take a few days; atropine does not cause any permanent damage to the body. A poisoner might choose to take a more direct approach to dispatching their victim, though, by administering a single large dose of atropine, perhaps in some food or drink. Due to its bitter taste, atropine is easily detected by anyone consuming tainted food, so the poisoner must ensure that a lethal dose is ingested in the first mouthful. This was the approach taken by the murderer in the Miss Marple short story *The Thumb Mark of St Peter.**

Even if the victim has ingested a fatal dose there is still a chance of recovery, as a range of antidotes and treatments is available. In a rare moment of compassion for her victims, Dame Agatha tells us of one antidote for atropine poisoning, pilocarpine, in *The Thumb Mark of St Peter*; Geoffrey Denman calls out 'pilocarpine' as he succumbs to the fatal effects of the atropine that had been added to a glass of water by his bedside. Christie was not kind enough to have Geoffrey Denman survive; witnesses at his death think he's delirious and talking about fish, with one witness saying it 'sounded like "pile of carp"'. The short-sighted doctor attending Geoffrey misses the tell-tale signs of dilated pupils.

A post-mortem examination of Denman's body fails to reveal the presence of atropine, though detection can be difficult if the post-mortem is delayed. Atropine leaves no obvious indications post-mortem; even the dilated pupils so characteristic of atropine use are an unreliable pointer after death, since the pupils naturally dilate as the muscles relax. In

---

*This short story is part of a collection entitled *The Thirteen Problems*. This was called *The Tuesday Club Murders* in the United States.

cases of suspected poisoning it would be normal to take samples from the stomach contents and body tissue to analyse for toxic compounds. Toxic metals are relatively easy to extract in such cases, as the surrounding tissue can be destroyed by the heat of chemical action, leaving the metal behind. Organic compounds such as atropine are generally destroyed by such processes, though, so they must be extracted and isolated intact before they can be identified. A method for the extraction of plant alkaloids has been available since 1850, when one was devised by Jean Stas (see page 169). A modified form of Stas's methodology is still in use today.

Once a plant poison has been isolated from human remains it has to be identified. In the past identification was often carried out by taste,[*] or by administering some of the extracted compound to a test animal such as a mouse or a frog to watch what symptoms the animal developed. If the symptoms of the animal matched the symptoms shown by the human victim the poisonous element had successfully been isolated. The symptoms of poisoning could then be used to identify what the chemical was.

Today a range of analytical techniques is available to confirm the presence and identity of a poison, if poisoning is suspected. Chromatography ('colour-writing') is a method for separating out components of a mixture. First developed in 1855, this method was initially used to separate out plant pigments. You may remember as a child putting dots of food colouring or coloured inks on pieces of paper and dipping the ends in water. The water slowly rises up the paper by capillary action, carrying the ink along with it. The varying solubility of the dyes and inks in water means they travel at different speeds up the paper, and the result is a series of differently coloured dots in a vertical line on the paper. The distances between the starting point and the point where the dots of ink stop are

---

[*]Hopefully in judiciously small quantities, to stop the pathologist from succumbing to the same poison. This is no longer considered a reliable method of identification.

characteristic for each ink. Scientists today use a variety of subtly different chromatography techniques and materials, but the principles are essentially the same. Samples extracted from tissues can be separated out by their differing solubilities in different solvents. Compounds that are very soluble are swept along in the solvent as it moves through the apparatus, while less soluble compounds lag behind. The rate of travel of compounds in an unknown sample is compared to known, pure samples. Identical rates under identical conditions indicate that the substances in the known and unknown samples are the same. This technique can also be used to determine how much of a plant alkaloid is present.

When Agatha Christie wrote *The Thirteen Problems* in 1932 chromatographic techniques were in their infancy, and commercial systems would not appear until after the Second World War. Chemical tests may have been available, and reactions between certain compounds carried out under certain conditions can produce characteristic colours, but these are unreliable (see page 194). Despite the problems faced by 1930s pathologists, it should have been possible to identify atropine in Geoffrey Denman's remains had they known to look for it. Miss Marple, however, had no need to rely on toxicology reports; she identified the poison from the slightly obscure 'pilocarpine' clue.

⌬⸫

Pilocarpine is another plant alkaloid, isolated from the leaves of jaborandi, *Pilocarpus pinnatifolius*. It binds to the same receptors as atropine but acts as an agonist, activating them to cause increased sweating and salivation, and slowing the heartbeat. Pilocarpine has medical uses in sweat tests, to measure the amount of chloride and sodium excreted in sweat, and to treat dry mouth, a condition often experienced after radiation therapy for head and neck cancers. It has also been used, in the form of eyedrops, to treat glaucoma. Because the effects of atropine and pilocarpine are opposite they can act as antidotes to each other.

Pilocarpine has been used in a real-life poisoning case, and atropine was administered by doctors who treated the victims. Unfortunately two of the victims, residents at a long-stay psychiatric establishment, West Park Hospital, did not survive. On 14 August 1985, some of the patients on the Exford ward at West Park were taken ill after their evening meal. There were bouts of coughing and excessive salivation, five of the 24 patients on the ward had difficulty breathing, and three had to be taken to the nearby Epsom District Hospital for treatment. Two women, 82-year-old Nora Swift and 99-year-old Florence Reeves, died despite the best efforts of the hospital staff. Urine samples taken from the women were analysed and showed the presence of pilocarpine and atropine. The atropine had been administered by the hospital to treat the symptoms displayed by the patients; it was the pilocarpine that had poisoned them.

Pilocarpine eyedrops were used on the Exford ward to treat several patients who suffered from glaucoma. All the medicines were stored in a room that was supposed to be locked, but in reality it rarely was. It was suggested that the drug had been added to the evening meal, but when everyone had finished eating the leftovers had been disposed of and the plates and cutlery washed up by the kitchen staff. There were only a few scraps of cottage pie left for the forensic team to analyse for traces of poison, but nothing was found. It was eventually determined that the pilocarpine had been added to the soup, because many of the patients had noticed a bitter taste and eaten very little of it. This accounted for the fact that not everyone was taken ill, and that most survived; just four drops of pilocarpine would be enough to kill an elderly patient.

The average age of the patients on the Exford ward was 88 and many were in an advanced stage of dementia; others had difficulty communicating because of their particular form of psychiatric disorder. The police struggled to question the patients, but it emerged that one woman had been upset when two others took her customary place at the dinner table. There were also disagreements among the staff that had resulted in one member being dismissed. Despite these tentative leads, whoever it was that added the poison to the soup, and the motive behind it, remains a mystery.

꒛ᆛᆺ

Pilocarpine has specialised medical applications and would not be part of most people's medicine cabinet, nor would it be carried by doctors on their rounds. Even if the doctor attending Geoffrey Denman in the Christie story had recognised the symptoms of atropine poisoning, it is unlikely that he would have had pilocarpine to hand. However, in cases of atropine poisoning other stimulants have been used, even caffeine.

As well as being an antidote for muscarin and pilocarpine poisoning, atropine is an antidote for poisonings from organophosphorus compounds. There are a huge number of compounds in this category of chemicals; they all contain phosphorus, carbon, hydrogen and oxygen atoms in various combinations and arrangements. These were initially developed in Germany in the 1930s as pesticides, but it was soon clear that some of them had serious effects on human health, including difficulty breathing and a dimming of vision as the pupils contracted to pinpoints. The Nazis went on to develop several of these compounds as chemical weapons; they were given names such as sarin, tabun and soman, but thankfully none were used during the Second World War.[*]

The best-known of these organophosphorus chemical weapons is sarin; sadly it *has* been used since the Second World War. Sarin is a relatively simple molecule that can be manufactured easily and stockpiled without its potency deteriorating over time, making it a popular choice for oppressive regimes and terrorist groups. Sarin interacts with cholinesterase, the enzyme that breaks up acetylcholine after it

[*]Mainly because Hitler couldn't believe that the Allies didn't have their own similar, but larger, stockpiles; the Allies did indeed have their own versions of organophosphorus-based chemical weapons, but again they were never used.

has done its job of activating muscarinic receptors (the ones that slow the heart rate, and increase the output of fluids like tears and sweat). If the acetylcholine molecules are not removed they continue to stimulate the receptors; the target organ then goes into spasm, eventually becoming paralysed, and death occurs because the muscles required for breathing stop working. Atropine opposes sarin poisoning; it blocks the muscarinic receptors without stimulating them, so the excess acetylcholine has only a limited effect.

Sarin was used in 1995 in attacks on the Tokyo subway by members of the Aum Shinrikyo doomsday sect. Thirteen people died, and several thousand required hospital treatment. The sect had previously carried out a smaller-scale attack, which alerted the authorities to its intentions. When the second attack was carried out doses of atropine were available, and many lives were saved because of this. Soldiers deployed in situations where they may encounter organophosphorus compounds, such as in the case of nerve-gas attack, carry doses of atropine with them.

Some organophosphorus compounds, organophosphates, also have a beneficial role and are widely used as pesticides in agriculture today. These compounds are considerably less toxic than those used as chemical weapons, and they are also less toxic than the organochlorine-based pesticides such as DDT that they replaced. Accidental and intentional poisonings with agricultural organophosphates do occur, and they have the same symptoms as the organophosphorus compounds of warfare. Fortunately atropine can again be used as an antidote in emergency situations, and these symptoms can be treated. However, studies are ongoing into the health effects of long-term exposure to low levels of organophosphate pesticides.

Some real-life cases
Real-life atropine murders are very rare, even though plants containing significant quantities of the poison grow wild in many countries. This is probably due to the ease of identifying and treating the symptoms. However, there have been some

atropine poisonings, and one bungled murder attempt may have been inspired by one of Agatha Christie's stories.

Atropine was used in a failed murder attempt in Scotland. In 1994 Alexandra Agutter was poisoned by her husband Paul, a biology lecturer at Edinburgh University. He had added atropine to a bottle of tonic water that he used to make his wife a gin and tonic. Alexandra complained of the bitter taste and only drank some of the tipple but she had ingested around 150mg, more than enough to kill her. She soon felt very ill, and five minutes later tried to stand but felt dizzy and collapsed on the floor. She had a pain in her throat, and started to hallucinate. Paul promised to ring for help but called their local doctor rather than the emergency services. The doctor was out, so he left an urgent message for him. When the doctor picked up the message he immediately rang for an ambulance and set out for the Agutters' home. When the doctor and the ambulance arrived it was obvious that Alexandra was gravely ill, and they suspected she had been poisoned. The ambulance man took possession of the drinks for analysis. Alexandra was rushed to hospital, and though she was very ill for some time, she recovered.

Although it was clear that Alexandra had been poisoned with atropine, no immediate suspicion fell on her husband. Paul Agutter had been very careful in his planning. He had chosen atropine for a number of reasons: he was well aware of its lethal properties, it was easy for him to obtain by stealing it from the research lab where he worked, and the bitter taste would, he thought, be disguised by the bitter taste of the tonic water. He also took great pains to leave a trail of red herrings that would divert attention from himself.

Agutter added atropine to a number of tonic-water bottles, not just the one he used to make his wife's drink. These bottles contained between 11mg and 74mg of atropine, not enough to kill, but enough to make anyone who drank them very ill. Paul left the bottles on the shelf of a local supermarket. His plan was to make the authorities believe that a psychopath was deliberately tampering with bottles of tonic water. But CCTV in the supermarket captured him in the store, and one of the employees remembered him placing bottles of tonic water on the shelves.

Eight people were subsequently admitted to hospital with atropine poisoning after drinking tonic water from the same shop. All the bottles of tonic water were taken from the shelves and tested; six more were found to contain atropine, and a nationwide alert asked people to return bottles of tonic water bought from the same supermarket chain. The story became front-page news, and Paul Agutter took part in a press conference with the police asking for help to find the culprit.

Agutter's mistake was failing to tidy up the evidence in his own home. The amount of atropine in the bottle of tonic water he used for Alexandra's drink was far higher (300mg) than the amounts he added to the other bottles. Had he substituted the bottle with the lethal dose for one intended for the supermarket and disposed of it before the ambulance arrived he might never have been suspected. Paul Agutter was found guilty of attempted murder and served seven years in prison.

✿✾✤

One of the most interesting – but now almost forgotten – cases of atropine poisoning occurred in 1977 in Créances, France. Roland Roussel, a 58-year-old office worker, plotted to kill a woman he thought responsible for the death of his mother. He added atropine from an eyedrop solution to a bottle of Côtes du Rhone wine that he left at his uncle Maxime Masseron's house, where the woman frequently visited. Roussel's uncle and aunt abstained from drinking alcohol outside of the holidays, but his intended victim was known to drink wine at the house. Unfortunately Uncle Maxime decided to keep the wine for a special occasion, and he opened the bottle on Christmas Day, pouring a glass for himself and his wife. Maxime died at the scene, but Mme Masseron (whose full name has unfortunately been lost in the mists of time) fell into a coma, and was rushed to hospital by neighbours.

It was thought that the couple had been victims of accidental food poisoning, and the police were not alerted until a few days later when a local carpenter and the victim's son-in-law went to the house to place Maxime's body in a coffin. Finding wine on the table from Christmas Day, both men decided to drink a glass, and they quickly became violently ill. Within an hour they had both fallen into a coma, but thanks to medical treatment their lives were saved. The police quickly turned their attention to Roland Roussel, and they found a copy of Agatha Christie's *The Thirteen Problems* in his apartment, along with other magazine and newspaper articles on poisons. A gendarme stated that 'I'm not saying Roussel was inspired by the book, but we found it in his apartment with the relevant passages on poisons underlined, and it was with that poison that the victim was killed.'

Agatha and atropine

In Agatha Christie's short story *The Cretan Bull*, Hugh Chandler displays all the symptoms of atropine poisoning, which a doctor would easily be able to recognise and treat. But instead of consulting a doctor, Diana Maberly, Hugh's fiancée, consults Hercule Poirot; fortunately for Hugh, Poirot's little grey cells are up to the challenge.

For the past year Hugh has been suffering from hallucinations, and a series of macabre incidents at his home mean drastic action had to be taken by Hugh's family. Hugh suffered from terrible, vivid dreams, and wandered around at night. To stop him or his family from coming to any harm Hugh was locked in his bedroom at nights, but sometimes he managed to escape. When he woke in the mornings he would find blood on his hands, and the bodies of slaughtered sheep were found in fields adjacent to the grounds of the house. Hugh was threatening to break off his engagement to Diana because he was worried for her safety.

When Poirot arrives on the scene he asks Hugh to describe his symptoms, which include a dry mouth and difficulty swallowing. Hugh also describes a feeling of flying, a sensation often experienced with atropine; belladonna and mandrake were the key ingredients in witches' flying salves because of the presence of atropine in these plants. These salves were prepared using fats to dissolve the atropine and increase its absorption into the bloodstream when the salve was rubbed onto the skin. The effect of atropine on the central nervous system, and specifically the brain, creates hallucinations that often take the form of dissociation; the feeling of the mind leaving the body gives a sensation of flying. Women accused of witchcraft often genuinely believed they had flown.

While Hugh talks to Poirot he describes a vivid hallucination of a skeletal figure standing nearby. Hugh is bewildered, terrified and clearly believes himself to be going mad, but Poirot has his doubts. Poirot believes the hallucinations and other symptoms Hugh is experiencing are due to deliberate poisoning with atropine, and he sets out to prove it.

Hugh had a shaving rash and was using a cream to soothe the skin. Poirot takes a sample of the cream, and analysis shows that it contains atropine sulfate, a compound that had been prescribed to Hugh's father for an eye condition. By having the prescription copied the poisoner could obtain the atropine sulfate and avoid awkward questions from the pharmacist. In 1947, when the book was written, this compound was likely to have been dispensed as a solution of 260mg of atropine sulfate dissolved in one ounce (approximately 30ml) of water. This would be enough to kill an adult man if given in a single dose, but the prescription would be expected to last for a month if used daily for both eyes. The daily dose would be approximately 4mg per drop, roughly the level at which toxic symptoms might start to appear. The atropine sulfate obtained by the poisoner in *The Cretan Bull* would have been mixed with the cream, which wouldn't have changed its appearance, apart from perhaps making it a little runny. Poirot suggests that the atropine sulfate was extracted from the medicine before being added to the cream. Extracting the atropine sulfate, or concentrating it, could easily be achieved by evaporating the water from the eyedrops to leave a solid residue. Face creams, or cold creams, are a mixture of oils in water, and if the poison was added as solid atropine sulfate it would dissolve in the water already present in the cream.

It is difficult to establish the dose Hugh must have been receiving, as we do not know how big the tub of cream was, or how much he used per day. Hugh must have been exposed to a larger daily dose than that designed to treat his father's eye condition in order to experience toxic effects. The amount of atropine sulfate absorbed through intact skin is usually very low, but the abrasive effect of shaving would have made the chemical far easier to absorb when the cream was applied to broken skin. As the poison took effect a rash appeared on Hugh's face, which would have allowed even more atropine sulfate to be absorbed – and probably caused him to apply yet more of the tainted face cream. Poirot's intervention and prevention of further poisoning saves Hugh's life; from what we know of this chemical, we can expect that he would go on to make a full recovery, and marry his betrothed, Miss Diana Maberley.

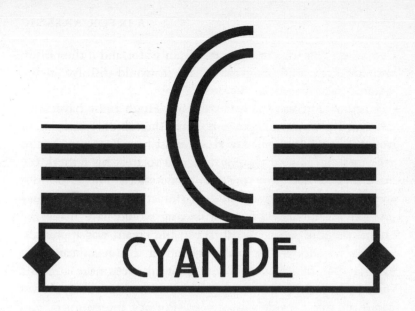

Sparkling Cyanide

The Strongest Poison ever known
Came from Caesar's Laurel Crown…

William Blake, *Auguries of Innocence*

Cyanide makes an appearance in no less than ten Agatha Christie novels and four short stories, where she uses it to bump off 18 characters. Christie had her murderers administer the poison in inventive and effective ways, including by injection, in drinks, in smelling salts and even in a cigarette. Her descriptions of the poison, the symptoms displayed by the victims and the possible sources of cyanide exhibit a high degree of accuracy. Rather than cataloguing the murders one after another, I'm going to focus on one book in particular, and that of course has to be *Sparkling Cyanide*.

Sparkling Cyanide, or *Remembered Death* in the United States, was written by Christie in 1945. The story centres on the well-to-do Barton family and a small group of friends, acquaintances and hangers-on. The novel opens with the recollection of events surrounding the dramatic death of Rosemary Barton at the Luxembourg restaurant by those who witnessed it. Seven people had dined together to celebrate a birthday; the lights went up after the cabaret, Barton took a sip from her champagne glass and then dropped dead, face down on the table, with her face blue and fingers twitching from convulsions. It was declared that death was from potassium cyanide poisoning, and the verdict was suicide.

Six months later, Rosemary's husband George Barton receives an anonymous note suggesting that Rosemary was murdered. Rather than do the sensible thing and tell the police, George embarks on an elaborate but insane plan to expose his wife's murderer. Exactly one year after the dreadful 'suicide', George gathers together the six diners who were present at the original party. George also hires an actress, made up to look like Rosemary, with the idea being to have her appear during the dinner and startle a confession out of the murderer. The plan fails spectacularly; when George drinks a toast to the memory of his late wife he suddenly turns purple and falls flat on the table. It takes one and a half minutes for him to die, his drink being laced with the same poison as his wife's, one year previously. Fortunately, George had revealed some of his suspicions and plans to his friend, Colonel Race, an intelligence officer, who then works with the police to solve the crime.

In 1945, many poisons were terrifyingly easy to obtain and few antidotes were available. Thankfully things have changed, but cyanide still has a reputation as a horrible and frighteningly effective poison – and with good reason.

The cyanide story
The name 'cyanide' comes from the Greek word *kyanos* meaning 'dark blue', but by quite a roundabout route. Prussian Blue ($[Fe_7(CN)_{18}]$), an intense blue pigment commonly used by artists, was the compound used in 1752 to produce hydrogen cyanide (HCN), or prussic acid as it was called, by the French

chemist Pierre Macquer (1718–1784). 'Cyanide' was therefore the name given to molecules containing a cyanide unit, even though very few of them are blue.

Cyanide is simply two atoms, one of carbon and one of nitrogen, bonded together as a unit (–CN) that forms part of a larger molecule. There are a staggering number of cyanide compounds, and they occur throughout the natural world as well as in synthetic versions. Their toxicity depends on the ease with which the bond between the cyanide unit and the rest of the molecule can be broken. For example, the bond between cyanide and a hydrogen atom in hydrogen cyanide (H–CN) is very easily broken, and the compound is therefore extremely toxic; 50–150mg can kill an adult. However, the same cyanide unit bonded to a methyl group to form methyl cyanide (CH_3–CN) is much less toxic (by a factor of approximately 5,000)[*] because the bond is more difficult to break. If accidentally ingested, most of the methyl cyanide would be excreted by the body long before the cyanide could be released.

Many plants contain cyanide compounds, but some are more dangerous than others. It depends on the type and amount of cyanide compound present. The seeds or pits of plants of the genus *Prunus* all contain cyanide; the stones or pips of peaches, cherries, apples and bitter almonds can be particularly dangerous and potentially lethal in large quantities. This is because the seeds contain cyanide in the form of a compound called amygdalin, which is readily metabolised by enzymes in the small intestine to release hydrogen cyanide.

Agatha Christie often refers to the scent of bitter almonds, to indicate that cyanide had been used. In fact, it's not that cyanide smells of bitter almonds; bitter almonds smell of cyanide.

Methods for distilling cyanide from natural sources have been known for thousands of years. One of the first references to poisoning dates from an ancient papyrus that refers to a

---

[*]Though still not good to drink or wash in.

lethal punishment for speaking the name of God: 'repeat not
the name I.O.A. [i.e. 'Jehovah'], under the penalty of the peach
tree';* indeed, in Agatha Christie's *Hallowe'en Party* the
murderer commits suicide by drinking a liquid smelling of
peaches. Laurel leaves are another natural source of cyanide
that has been used since the time of the Romans, giving
'Caesar's Laurel Crown' in William Blake's poem a particularly
sinister symbolism.

Natural sources of cyanide have also been used in more
recent times. In 1845 apple pips were used as a defence in the
trial of John Tawell, who was accused of murdering his mistress
Sarah Hart with cyanide.† Tawell was known to have purchased
prussic acid from a pharmacy some time before Hart died.
Prussic acid was known to be a dangerous substance but Tawell
claimed the prussic acid was for 'external application'. At the
time there was a vogue for using solutions of prussic acid as a
skin lotion. One commercial product, 'Dr Eliotsom's Lotion of
Prussic Acid', was recommended for moistening the skin
before and after shaving. Lethal amounts of prussic acid can be
absorbed through unbroken skin; shaving cuts and abrasions
could only have eased the absorption. Thankfully the fashion

---

*Books on poisons and toxicology often state that the ancient
Egyptians used cyanide from peach stones as a poison. A paper
published in 1938 traced these statements back to an alleged
translation by Duteil of a passage in 'an extremely ancient papyrus in
the Louvre', which is as follows: *Ne prononcez pas le nom de IAO, sous
la peine du pecher* ('Speak not the name of IAO [Hebrew shorthand
for the name of God] under the penalty of the peach-tree'). This first
appeared in an 1842 book on the history of chemistry by F. Hoefer.
He claimed to be quoting Duteil but the quote has not been found
in Duteil's writing. In 1938 the Louvre had four documents that
might be expected to contain such a quote, the 'demotic magical
papyrus' and three 'Greek magical papyri' but they contain no such
reference to the penalty of the peach. The source of the quote
remains a mystery.

†Tawell lived in London but kept his mistress in a cottage in Salt Hill,
near Slough, Berkshire.

did not last. Tawell's purchase was added to a bottle of beer, which Sarah drank. A neighbour saw Tawell leave the house, and hearing Sarah's cries went to see if she was all right. She found Sarah writhing on the floor in agony, and frothing at the mouth. She died before the doctor arrived.

The police were alerted and raced after Tawell, but they couldn't stop him before he boarded a train for London. A telegraph message was sent to London describing him and instructing the police there to arrest him. It was the first time the telegraph system had been used in this way, and the case generated huge publicity because of it.

At Tawell's trial, the defence barrister, Sir Fitzroy Kelly, proposed that the cyanide that had killed Sarah had come from the pips of the apples she was so fond of eating. The lethal dose of apple pips is about 200g; Sarah would have had to eat thousands of apples to ingest enough pips, and they would also have had to be well chewed to release the poison. The jury were unimpressed, and he was found guilty. Tawell was executed for his crime and his barrister was known as 'Apple-pips' Kelly for the rest of his career.

Bitter almonds and apricot kernels contain considerably more cyanide-containing amygdalin than apple pips; just a few apricot kernels can be deadly, but there is another plant that contains even more cyanide. Cassava is probably the most dangerous cyanide-containing plant regularly consumed by humans. It forms an important part of the diet of millions of people living in the tropics. Bitter cassava can contain 1g of cyanide in every kilogram of root, in the form of two compounds, linamarin and lotaustralin, which share a close similarity with amygdalin (all three are in the same class of compounds, the cyanogenic* glucosides). Sweet cassava is a different variety that contains considerably less cyanide, but people tend to grow the bitter variety as it is more resistant to pests and, quite possibly, to thieves. Eating a few mouthfuls of

---

*'Cyanogenic' simply means something capable of generating hydrogen cyanide.

raw bitter cassava is unlikely to kill you straight away, but it will cause severe illness. Raw cassava must be processed to remove the cyanide by grinding the vegetable into flour and soaking it in water for anything between five hours and three days (variations depend on cassava variety and local tradition). The soaking process allows the enzyme luminase, which is also present in the vegetable, to convert the cyanogenic glucosides into hydrogen cyanide, which evaporates from the water. The amount of cyanide compounds in the root increases in droughts, and with less water available for the soaking process people are particularly vulnerable to the effects of the poison at these times. Failure to prepare cassava properly can lead to goitre and damage to the nervous system, manifesting itself as a condition known as *konzo*. Individuals with *konzo* have problems walking, and with coordination generally. The condition is irreversible, and can be fatal.

Eating very small quantities of cyanide poses no health risks: we have all swallowed the occasional apple pip with no adverse reaction, because the body has some immunity against cyanide compounds.* Humans evolved on a diet featuring foraged plants, and we have adapted over thousands of generations to cope with a certain level of cyanide exposure. Almost every cell inside the body contains an enzyme called rhodanase that converts cyanide (–CN) into thiocyanate (–SCN). Thiocyanate is about a thousand times less toxic than cyanide, and it can be readily excreted in urine. Our bodies have the capacity to process around 1g of cyanide every 24 hours; problems occur when the system is overloaded by a sudden influx of large amounts.

One animal species, however, is completely immune to the effects of cyanide. The greater bamboo lemur has evolved a taste for the shoots of the Madagascan giant bamboo, and it eats almost nothing else. These shoots are laced with cyanide

---

*The tough seed coating will also prevent the release of much of the poison.

compounds, but the lemur has evolved an immunity that allows it to feast away to its heart's content.

There are many other sources of cyanide. As Agatha Christie mentions in *Sparkling Cyanide*, one legitimate reason for having cyanide compounds in the home, aside from those found within fruits and nuts, is for photography. The compounds involved are potassium ferricyanide ($K_3Fe(CN)_6$), an orange-red crystalline solid; and a compound we've already encountered, Prussian Blue, known in photographic circles as ferric ferricyanide($[Fe_7(CN)_{18}]$). These compounds are used to alter the tones of prints and to produce cyanotypes or blueprints. Neither is particularly toxic, but both can generate hydrogen cyanide if mixed with acid.

Another form of cyanide compound, and a form commonly associated with murder and suicide, is a cyanide salt such as potassium (KCN) or sodium (NaCN) cyanide. Like table salt, these compounds dissolve easily in water to release the cyanide unit. The water molecules react with the cyanide salt to form hydrogen cyanide, in a process called hydrolysis. The bonds between cyanide and the rest of the molecule are broken easily; therefore potassium and sodium cyanide are both very toxic, with a lethal dose for an adult being around 200–300mg. These salts have applications in gold mining, as potassium cyanide reacts with gold to form soluble gold cyanide compounds that can be washed out of rocks and collected. The gold can then be easily extracted from its cyanide compound. Cyanide salts remain industrially important chemicals, and they can often be found in locked cupboards inside chemistry research labs. Today their sale is tightly regulated and use is restricted to those who know what they are doing.

Potassium and sodium cyanide were once commonly used as insecticides. In fact one suspect in *Sparkling Cyanide* is questioned about the contents of his gardener's shed, and another character discusses the high number of wasps' nests they had that summer. A small quantity of the cyanide salt was shaken in a bottle of water or weak acid to release hydrogen cyanide, which would kill the wasps or other insects as well as anyone foolish enough to breathe in the gas. A similar process was used in the gas chambers inside US prisons, where convicted criminals were executed. The first execution by gas chamber was carried out in Nevada in 1921. It was supposed to be a quick death, but some prisoners held their breath and struggled. The prisoner was taken into an airtight room; after the door had been sealed, a lever was pulled that dropped sodium cyanide pellets into a bucket of sulfuric acid under the prisoner's chair. After the prisoner died – sometimes more than eight minutes after first exhibiting convulsions – the room was purged with air. The last execution by gas chamber was in 1999. Lethal injection is now the preferred method of execution in the United States.

The use of cyanide to murder people was perfected by the Nazis during the Second World War; cyanide compounds were used to kill millions in the Holocaust. Using the excuse of pest control, the Nazis manufactured and transported tonnes of Zyklon B (a trade name for a cyanide-based pesticide) to the concentration camps. Tins of Zyklon B contained hydrogen cyanide, a stabiliser and an odorant (ethyl bromoacetate), the latter presumably acting as a warning for the guards in case there was a leak. Hydrogen cyanide boils at $26°C$ and would quickly vaporise in the hot confines of the gas chambers. In large doses hydrogen cyanide is mercifully quick-acting and it can kill almost instantly. As the war drew to a close, many of the Nazi leaders, including Hitler himself, bit into cyanide capsules to kill themselves rather than face capture and trial.

How cyanide kills
Cyanide kills owing to its interaction with a specific enzyme, cytochrome c oxidase. Regardless of whether cyanide is introduced to the body in compounds such as amygdalin or in the form of cyanide salts, the result is the same. Enzymes in the gut interact with cyanogenic glucosides, and cyanide salts react with stomach acid; in both cases hydrogen cyanide is

produced. Hydrogen cyanide is rapidly absorbed into the bloodstream and transported to the places where it does the real damage.

In the bloodstream, cyanide attaches to haemoglobin, the protein that carries oxygen from the lungs to the rest of the body. Each haemoglobin protein contains four globular subunits, each containing a single atom of iron to which oxygen – or cyanide – binds. Cyanide binds more strongly to iron, so it can displace the oxygen molecules. Haemoglobin represents a highly efficient system for distributing oxygen around the body; this same efficiency allows cyanide to be rapidly delivered to the sites where it does the most damage – inside our cells.

Almost every cell in our body contains structures called mitochondria, which function as the 'engines' of the cell by carrying out respiration. Respiration is the process whereby oxygen from the lungs, delivered by haemoglobin, reacts with glucose in a series of controlled steps to release energy in the form of adenosine triphosphate (ATP). Cells with high energy demands have large numbers of mitrochondria. Liver cells can contain more than 2,000 mitochondria (while red blood cells have none). Mitochondria are particularly important in heart and nerve cells owing to their high energy demands. Each step in the complex process of energy release and ATP production is controlled by a specific enzyme. The cytochrome c oxidase enzyme is the final step in the cascade of reactions of respiration. An atom of iron lies at the active site of cytochrome c oxidase, and it is here that an oxygen molecule normally binds. As in haemoglobin, cyanide readily takes the place of oxygen; it binds to the iron atom irreversibly, stopping the flow of chemical reactions in their tracks. With the source of its energy blocked, the cell rapidly ceases to function, and cell death occurs quickly.

Swallowing large doses of cyanide can kill in minutes owing to massive, widespread cell death; though some individuals may survive longer, death usually occurs within four hours. The symptoms shown by cyanide victims in the short window of time before they die can include dizziness, rapid breathing, vomiting, flushing, drowsiness, a rapid pulse and unconsciousness.

Is there an antidote?

Any emergency treatment has to be given quickly, to prevent the cyanide reaching the cytochrome c oxidase enzyme. The problem with any antidote, though, is the speed at which cyanide acts. Even today, with a range of antidotes available, 95 per cent of accidental cyanide poisonings are fatal. Mouth-to-mouth resuscitation is not advisable when dealing with cyanide poisoning, as the rescuer is likely to inhale hydrogen cyanide from the stomach or lungs and be poisoned themselves. Today, people working with cyanide compounds as part of their job will usually have a cyanide antidote kit to hand in case the worst happens.

The first known effective antidote was amyl nitrite, a compound that was first synthesised in 1857. Its effect on the human body was soon noticed, and by 1859 the compound had been shown to relax smooth muscle and was being used to treat heart pain and angina. Around the turn of the century its effectiveness in the treatment of cyanide poisoning was also noted; the results of a study carried out in the United States and published in 1933 confirmed this. If UK doctors had read the paper and adopted its recommendations this treatment might have been available to George and Rosemary Barton in *Sparkling Cyanide*.

Amyl nitrate is a clear, colourless liquid that boils at 21°C. Glass vials would be popped open so the vaporised liquid could be inhaled, hence the name 'poppers' for modern recreational use of the drug. One of the actions of the compound is the conversion of haemoglobin to a similar compound, methaemoglobin; cyanide binds preferentially to the iron in the methaemoglobin rather than the iron in cytochrome c oxidase. The resulting compound is not toxic and is safely excreted in urine, leaving the cytochrome c oxidase uncontaminated and the patient able to process oxygen normally. This treatment is still used for cyanide poisoning today, but it does have a downside.

Methaemoglobin is unable to bind with oxygen, so the body suffers a reduced oxygen capacity. This can lead to symptoms including headaches and convulsions. Another chemical, methylene blue, is needed to convert the methaemoglobin back to haemoglobin, so it can function normally again.

Today there are many antidotes for cyanide poisoning, but none of them is completely free from complications. Many work in a similar way to amyl nitrite by providing an alternative site for the cyanide to bind to, so it doesn't affect cytochrome c oxidase. One example of this is the use of hydroxocobalamin, a form of vitamin B12. It acts in a similar way to methaemoglobin, and the resulting cyano-complex is non-toxic and excreted in the urine. This antidote has the added benefit of leaving all the body's haemoglobin untouched so no further treatment is required. Unfortunately, hydroxocobalamin is expensive, and not universally available. A cheaper alternative is dicobalt–EDTA, sold as Kelocyanor. Cyanide binds to cobalt just as well as it does to iron, but cobalt compounds are themselves toxic. A patient given Kelocyanor when they have not been poisoned by cyanide may die of the cure.

A different approach makes use of the body's own defence against cyanide, the enzyme rhodenase, which evolved over the millennia to deal with cyanide in our diet. The enzyme uses thiosulfate ($S_2O_3^{2-}$) to convert cyanide (–CN) to thiocyanate (–SCN) but it works slowly, too slowly to be effective against a sudden large dose of cyanide. By supplying the body with extra thiosulfate, the enzyme can react with more cyanide. This treatment is slow-acting so it is often given along with amyl nitrite to speed things up. So far this method of treatment is based only on animal experiments, with few case studies. Giving oxygen is another treatment that can support life while the body metabolises the cyanide naturally, but it is not an antidote in itself.

Some real-life cases
Agatha Christie did her research on cyanide poisoning, but, excluding the use of hydrogen cyanide in the Holocaust, she

had quite a few real-life murder cases to draw on. Perhaps one of the best-known cyanide poisonings is a failed but famous attempt from 1916.

Confidant of the Tsarina of Russia, Grigori Yefimovich Rasputin, known as the 'mad monk', had more than a few enemies. Some of them – Prince Felix Yusupov, the Grand Duke Dmitri Pavlovich, and the right-wing politician Vladimir Purishkevich – apparently lured Rasputin to the Yusupovs' Moika Palace for cake and Madeira wine. The cake and wine were said to be laced with enough cyanide to kill 'a monastery of monks' but it left Rasputin unaffected. He was then shot, at least twice, but was still alive and fighting back against his would-be assassins. At this point he was beaten into submission, tied up in a carpet and dropped into the frozen Neva river. His body was recovered two days later, and a post-mortem revealed that he had died from drowning.

There are a number of theories that might explain what happened that day:

1. His assassins were terrible poisoners and did not put enough cyanide in the food to kill him, or mistook an innocuous substance for cyanide salts.

2. Rasputin suffered from alcoholic gastritis.

3. Suspecting someone might try and poison him, Rasputin dosed himself regularly with small amounts of poison to build up an immunity to a larger, normally lethal, dose.

4. The sugary cakes and wine acted as an antidote to the cyanide.

5. The story is made up and Rasputin was killed by a single shot to the head fired by a British secret service agent.

The first theory is now impossible to prove or disprove. The cake and wine were not analysed, and those involved changed their stories several times, making their testimony rather unreliable.

The second theory is certainly reasonable and based on good science. Alcoholic gastritis could offer some protection against cyanide poisoning because it causes a thickening or inflammation of the stomach lining, and a decrease in production of stomach acid. With less stomach acid about, less of the potassium cyanide would have been converted into lethal hydrogen cyanide. However, we do not know if Rasputin suffered from this condition; his sister reportedly claimed he suffered from an *excess* of stomach acid, and was therefore unlikely to eat cakes and wine in the first place.

The third theory is also worth closer scrutiny, as stories of Mithridatism, as it is properly known, have been around for at least 2,000 years. The King of Pontus, Mithridates, fearful of poisoners, apparently built up an immunity to poison by regularly administering sub-lethal doses over a long period of time. The concoction he developed included more than 50 ingredients and was said to protect against every known poison. When the King was captured, and wished to kill himself by poisoning, the attempt failed (obviously) and he had to ask a guard to run him through with a sword.

Much of Mithridates' life is the stuff of legend, but could someone develop an immunity to poison using this method, or an adapted version of it? The answer is yes – and no. It is possible to develop immunity to some animal venom by administering sub-lethal doses. This is used successfully by people who work closely with venomous animals, zoo-keepers for example. But animal venom is very different from cyanide; your body will not build up an immunity by eating small amounts of cyanide salts. It will process the cyanide into thiocyanate and get rid of it, or you will be poisoned and quite possibly die.

The fourth theory, the possibility of a glucose antidote, is also promising. Research on rats has revealed that glucose offers some protection against the effects of cyanide poisoning, though the mechanism is not known. One theory is that the cyanide reacts with glucose to form non-toxic compounds that can be excreted, but more work is needed to establish the exact method; glucose is not officially recognised as an antidote to cyanide poisoning.

The fifth Rasputin theory is, of course, the most likely explanation.

Christie and cyanide

The two deaths in *Sparkling Cyanide* are very similar. Both Rosemary and George Barton had lethal doses of potassium cyanide added to their champagne glasses. Potassium and sodium cyanide salts are white crystalline solids much like sugar or table salt in appearance. These salts can have a faint but characteristic smell of almonds, due to a reaction between the salt and moisture in the air producing small amounts of hydrogen cyanide. A small amount of white crystals that looked like sugar certainly would not have seemed out of place in a restaurant, and could easily be slipped into a drink without anyone suspecting. Champagne is mostly water and would have easily dissolved the potassium cyanide; champagne is slightly acidic (around pH4) and the acidity could have made the reaction to produce hydrogen cyanide faster, and made the smell and taste of bitter almonds stronger.

However, between 20 and 60 per cent of people *cannot smell cyanide*. The research done in this area is not huge – asking people to sniff cyanide, even in safe amounts, does not have volunteers queuing out of the door. Experimentation was mostly carried out in the 1950s and 1960s, though the phenomenon had been known about long before then. The conclusions of these studies vary, but the general consensus is that while there is a genetic component involved, previous exposure to cyanide in the environment seems to be the more important factor in determining an individual's ability to detect cyanide. Rosemary and George may not have been able to notice that their champagne had an unusual smell and flavour, and Christie made no comment on the characteristic cyanide smell on this occasion.

The effects of the poison took hold immediately after the champagne had been swallowed. Symptoms of cyanide poisoning generally appear between one and fifteen minutes after exposure; the quickest way to get the poison to take effect would be to introduce it directly into the bloodstream by injection. Inhalation of hydrogen cyanide gas would not take much longer. Ingested cyanide compounds would be expected to act a little more slowly, as there is a slight time-lag while the cyanide is absorbed through the walls of the gastrointestinal tract and makes its way into the bloodstream. Cyanide ingested on a full stomach would take even longer; Christie does not mention at what stage of the meal the poisoning takes place in *Sparkling Cyanide*.

The symptoms Rosemary and George displayed, gasping for breath and convulsions, are an accurate description of what you might expect to see in a cyanide victim. Christie makes a point of the blue colour of the victims, and it is suggested that this is a symptom of the poisoning. Unfortunately, and very unusually for Christie, she got this wrong. The blue colour, or 'cyanosed face' as Christie describes it, is a condition called cyanosis.

The word 'cyanosis' is also derived from the Greek word *kyanos*. It is caused by a lack of oxygen circulating in the body. When an oxygen molecule binds to the iron atom in haemoglobin, forming oxyhaemoglobin, it becomes bright red. Deoxyhaemoglobin is formed after the oxygen has been released; its colour is closer to the blue end of the spectrum.* An excess of blue-coloured deoxyhaemoglobin causes a noticeable blue colour in the skin. The condition can be localised owing to the cold (think of blue fingertips in winter), or because of a blocked artery, but it can also give a general blue colour to the whole body because of problems absorbing oxygen from the lungs. There are a variety of root causes for this condition, but none of them involve cyanide.

---

*Hence we see blue veins under our skin where deoxygenated blood is being transported back to the lungs to collect more oxygen.

Cyanide victims sometimes actually seem flushed in appearance, with a pink coloration to their skin. Cyanide-haemoglobin complexes are pink in colour, but this is not the only cause of the flushed appearance. Because oxygen is blocked from binding to cytochrome c oxidase, it is not consumed in the normal respiration process and its concentration in the blood builds up. With high levels of oxygen present the oxyhaemoglobin fails to release its oxygen, and red oxygenated blood returns to the heart and lungs through the veins. Hence the red flushing.[*]

Both victims in *Sparkling Cyanide* die very quickly; George takes one and a half minutes to die, and this would be plausible if he had been given a large dose of potassium cyanide. The potassium cyanide was supposedly contained within a 'cachet faivre'; aspirin and other remedies in powdered form were often folded inside slips of paper, a common form of packaging in the 1940s, and headache powders are still sometimes sold like this today. The powders are dissolved in water before swallowing; a normal dose of aspirin would be approximately 600mg, so a lethal (200–300mg) dose of potassium cyanide would easily fit in the cachet, and even leave room for more than was strictly needed.

The cause of Rosemary and George Barton's deaths was never in doubt. Few poisons act as quickly as cyanide, and traces of potassium cyanide were found in the champagne glasses as well as in an empty Cachet Faivre paper found under the table. A post-mortem examination would probably show indications of cyanide poisoning. Potassium and sodium cyanide are slightly corrosive, and may leave traces of burning on the lips and tongue. In the stomach the corrosive properties may have shown themselves by eroding the stomach wall, leaving behind a characteristic blackened discoloration for the pathologist to find. Hydrogen cyanide doesn't corrode the

---

[*]To be fair to Christie, I have read many different descriptions of the colour of corpses in cyanide poisonings. These range from flushed to ashen, though not blue.

stomach in the same way, but there are other signs to look for. There may have been an aroma of bitter almonds emanating from the viscera for the pathologist to smell, if he was able to, to confirm the presence of cyanide (inhaling cyanide fumes from cadavers can be a considerable hazard for mortuary staff). The cherry-red appearance of the blood would also indicate cyanide, although carbon monoxide poisoning can cause a similar coloration. But while detecting cyanide poisoning is relatively straightforward, determining the quantity ingested is much more difficult, even today.

Levels of both cyanide and thiocyanate in the body can be determined accurately, but there may be several natural or environmental sources of cyanide to complicate things. Cyanide is a common chemical unit, and cyanide compounds can be introduced into the body through the food we eat. Agatha Christie helpfully describes what was on the menu at the Luxembourg the night that Rosemary dies – Oysters, Clear Soup, Sole Luxembourg, Grouse, Poires Hélène (pears in a sugar syrup) and Chicken Livers in Bacon. George, in his attempt to expose the murderer at his recreation of the scene one year on, even replicates the food served to the guests. But there are no dishes on the menu that would deliver particularly high doses of cyanide.

However, there may have been a possible antidote amongst the comestibles, in the form of sugar. Champagne contains a little sugar, but the most sugary champagne contains only 50g of sugar per litre, compared to 150g per litre in the Madeira wine that Rasputin drank on his final night. But there were other sources of glucose at the Luxembourg. The sugar syrup in the Poires Hélène dish could have protected Rosemary and George against the worst effects of cyanide poisoning. Perhaps they died before they got to the dessert course.

Yet another source of cyanide, one that could have affected post-mortem analysis, is in cigarette smoke. Smoking was much more common in 1945 than it is today, and though Christie doesn't mention whether Rosemary or George were smokers, it would have been quite likely that they were.

Hydrogen cyanide is a common by-product of the combustion of natural materials such as tobacco, silk or wool, and some plastics also release hydrogen cyanide when they burn. Cyanide poisoning is thought to contribute to a significant proportion of deaths from smoke inhalation in fires, though cyanide levels are not always specifically tested in the remains of these victims, but it is an important consideration for fire-fighters.

In addition to all of these sources, cyanide compounds can also be produced in the body by the normal decay processes that occur after death. So the post-mortem picture can be complicated by many potential sources of cyanide, some of which may have been metabolised into thiocyanate before death, and the addition of new cyanide compounds afterwards.

The best thing the pathologist in *Sparkling Cyanide* could have done would have been to analyse the residues left in the champagne glass. If there was enough left in the glass, the concentration of cyanide, and therefore the total dose, could have been determined. This is still the best way of determining the dose in cyanide poisoning cases today, as unpicking the amount of cyanide from environmental sources and working out the dose from post-mortem blood levels of cyanide and thiocyanate is a task full of potential errors.

Little seems to have been done for Rosemary or George to try to save or revive them. There is no mention of emergency treatment, or even of calling an ambulance to rush them to hospital. Antidotes were available in 1945 – but that would not have made for a good murder-mystery novel.

DIGITALIS

Appointment with Death

'Winston, if you were my husband, I would flavour your tea with poison.'
'Madam, if I were your husband, I would drink it.'
Lady Nancy Astor and Winston Churchill

The quote above is typical of the acid-tongued exchanges between Lady Nancy Astor, the first female Member of Parliament, and Winston Churchill. Agatha Christie may have been thinking of Lady Astor when she created the character Lady Mary Westholme for her 1938 novel *Appointment with Death*. The two certainly bear a striking resemblance; Christie claimed, however, that her inspiration for this loud and opinionated character came from two women she had met in the Far East. Another character in the novel, Dr Gerard, comments thus on Lady Westholme: 'that woman should be poisoned … It is incredible to me that she has had a husband for many years and that he has not already done so.' But it turns out not to be Lady Westholme who is poisoned; the victim is another domineering woman, Mrs Boynton.

Appointment with Death is set in Jordan, where a tourist group visits the abandoned city of Petra. The tourist party is made up of the Boynton family, under the control of malicious Mrs Boynton; Miss Pierce, a timid former nursery governess; Dr Gerard, a psychologist;* the outspoken lady politician Lady Westholme; and Sarah King, a young doctor. On the first afternoon at Petra the group takes the opportunity to explore the site, leaving Mrs Boynton alone in the hot sun. When the party returns to camp Mrs Boynton is dead. The death would have been attributed to natural causes were it not for a mark left by a hypodermic syringe on Mrs Boynton's wrist, and some missing heart medicine. Fortunately for all concerned (though maybe not for Mrs Boynton), Hercule Poirot is holidaying nearby, and he is asked to investigate the true cause of death.

The suspected poison in the case of Mrs Boynton is digitalis, an extract of the foxglove plant, which is often prescribed to treat certain heart conditions. Digitalis is an effective medium of murder, with the added benefit that symptoms of an overdose resemble those of the disease the drug is prescribed for. Plus it is readily available, and lethal in very small quantities. Incredibly, very few murderers have used this toxic substance, or perhaps many have but they got away with it. But before you rush to take out hefty life-insurance policies on your closest and wealthiest relatives, or start growing foxgloves in your garden, remember that the drug is detectable even in minute quantities.

The digitalis story
'Digitalis' refers to a group of related compounds extracted from plants of the genus *Digitalis*, commonly known as

*Clinical psychologists and psychiatrists are required to have medical training. This proves to be very handy in this book.

foxgloves. There are more than twenty species, all containing digitalis compounds in varying amounts and proportions. In the plant, digitalis compounds act as a deterrent against animals that try to eat it. In mammals these compounds have a specific and dramatic effect on the heart. The chemical structure of these compounds includes a component known as a glycoside, hence digitalis compounds are known as 'cardiac glycosides'. Toxic digitalis compounds are found in all parts of the plant; they can irritate the skin, and cause delirium, tremors, convulsions, headaches and fatal heart problems if ingested.

Digitalis plants are native to western Europe, western and central Asia, north-western Africa and Australasia. Though the plants grow wild, they are often cultivated because of their spectacular spikes of flowers in a range of colours. In their first year foxglove plants produce just a stem and soft, hairy, lance-shaped leaves. In two Agatha Christie stories, the short story *The Herb of Death* and the novel *Postern of Fate*, foxglove leaves are used to poison a meal by mixing them with sage and spinach leaves. One murderer deliberately planted foxgloves in a kitchen garden amongst the sage so the leaves would be picked by accident (this seems an unlikely mistake for an experienced cook to make). The distinctive flowers appear in the second year of growth, and look like caps or bells.

The name 'foxglove' has been in use for hundreds of years, since at least the fourteenth century, but the origins of the name are obscure and many theories have been put forward to explain it. An attempt was made to throw some light on the subject by Dr Prior, an authority on the origin of popular names, in the 1866 book *English Botany*:

> *Its Norwegian name,* Revbielde, *foxbell, is the only foreign one that alludes to that animal ... In France it is called* Gants de Notre Dame; *in Germany* Fingerhut. *It seems most probable that the name was, in the first place, foxes' glew, or music, in reference to that favourite instrument of an earlier time, a ring of bells hung on an arched support ...*

Prior proposed an alternative theory in the same book. 'The "folks" of our ancestors were the "fairies", and nothing was more likely than that the pretty coloured bells of the plants would be designated "Folksgloves", afterwards "Foxglove".' The foxglove plant has been part of folklore for centuries and traditional medicine for just as long, where it has been used to treat heart conditions and dropsy. The first systematic and scientific study of extracts of the plant was not made until the late eighteenth century by William Withering (1741–1799), a doctor from Shropshire. Withering noticed that one of his patients suffering from dropsy recovered after using a herbal remedy given to him by 'the old woman of Shropshire'.

Dropsy, now known as oedema, is swelling caused by an accumulation of fluid in the body. Oedema has several causes, but is often due to either weak action of the heart or cirrhosis of the liver. Fluid filters out of the blood and is reabsorbed at the capillaries. The balance between filtering and reabsorption depends on the resistance of the blood and blood pressure. Under normal conditions the rate of filtration is higher than that of absorption, and the excess fluid is removed from the tissues by the lymphatic system. Fluid is returned to the blood from the lymphatic system at the superior vena cava, one of the large veins that take deoxygenated blood back to the heart. Fluid is permanently removed from the blood by filtration through the kidneys. All the blood in a human is filtered by the kidneys roughly every half-hour, and excess water is excreted via the bladder.

A person with a weak pulse due to heart failure is unable to expel blood from the ventricles of the heart efficiently. This results in an increase in blood pressure, and therefore an increase in the filtering of fluid into the tissues. The kidneys respond by retaining more fluid, so there is very little urine output. This leads to swelling in the legs and arms, and difficulty breathing as fluid builds up around the lungs.

After observing the successful treatment of his patient, William Withering sought out 'the old woman of Shropshire' and asked her what went into her herbal treatment.

She wouldn't disclose the recipe, but Withering persuaded her to give him some of the preparations. By examining them under a microscope he identified fragments of foxglove.

Withering began what were effectively clinical trials of digitalis, which ultimately involved 163 patients. He gave his dropsy patients small amounts of various foxglove preparations, and observed their effects. He found dried and powdered foxglove leaves given by mouth to be the most effective treatment. Carefully increasing the dose, he monitored his patients' progress, recording results, both successful and unsuccessful, in his notebook.

The principles behind clinical trials have ancient origins stretching back to the time of the Old Testament, but they were not routinely used to assess the positive and negative effects of diet or drugs until the twentieth century. Withering's detailed study stands out because of his systematic approach to the use of foxglove preparations, and his careful notation of the adverse as well as the beneficial effects of these preparations. In his investigation, Withering noted that foxglove was extremely effective for some of his patients but not others. We now know that herbal preparations would be effective for dropsy caused by heart conditions, but they offered no help to those who were suffering from dropsy caused by cirrhosis of the liver. Withering also noted that at higher doses toxic effects developed, and he described the symptoms in detail.

Withering wrote up his observations in a treatise entitled *An Account of the Foxglove and some of its Medical Uses* (1785), still considered a classic of its kind. The book was widely read, and treatments were given to an increasing number of patients. However, some doctors grew impatient with the cautious and slow approach to foxglove treatments. Withering recommended that very small doses should be given initially and slowly increased until the desired effect was observed, in order to prevent toxic side effects. The extreme potency of the compounds within foxglove meant that it was exceptionally dangerous, and it was very easy to give an overdose. Even a 'raw' dose (without purification of the active ingredients) will kill in quantities as low as 2g. Toxicity can also build up over time, as some digitalis compounds have a long half-life within the body.

Despite the known toxicity of foxglove preparations, physicians continued to experiment with them until after Withering's death, when digitalis fell out of favour. When Withering himself became ill his friends commented that 'The flower of physic is indeed Withering'.* He died in 1799 of consumption (i.e. tuberculosis), and a foxglove was carved on his tombstone. Interest in digitalis was only revived in the early 1900s.

Withering used raw foxglove leaf in his treatments, but with developments in chemistry it became possible to refine the drug into a blend of several cardiac glycosides extracted from the plant. Even so, many of these compounds were inactive at best, and possibly harmful. In *Appointment with Death*, Agatha Christie lists four active principles of the foxglove – digitalin, digitonin, digitalein and digitoxin – a bit of a mouthful, admittedly. In her time these compounds were not easy to isolate as they are relatively fragile and could decompose during the extraction processes that were used in the late nineteenth and early twentieth centuries. They are also quite large molecules and their precise structures were not elucidated until years after their isolation (for example, digitoxin was identified in 1875 but its structure wasn't determined until 1962).

Of the compounds listed by Christie, digitonin is now known not to be a cardiac glycoside, because it has no effect on the heart (it does cause the breakdown of red blood cells, though). Digitonin is a saponin, meaning it forms a soap-like foam when shaken with water. Digitalein is something of an unknown; the word fell out of use in the scientific literature around 1921. It was probably not a single pure compound; as

---

*'Physic' meaning 'medicine' in this case – an obsolete usage of the word. This does, though, explain why doctors are sometimes referred to as 'physicians'.

scientists painstakingly isolated and identified more compounds from foxgloves, the name digitalein became irrelevant. The other two, digitoxin and digitalin, are still prescribed today. Digitalin is now known as digoxin, and is the more potent of the two (being between ten and twenty times more effective).

Digitalis compounds are complex and difficult to synthesise in the laboratory, so the drug is still obtained from the natural source by extraction from the plant *Digitalis purpurea*, the purple foxglove that grows wild in Britain, which contains both digoxin and digitoxin.* Digitoxin is now rarely prescribed as it has a considerably longer half-life in the body, six days as opposed to 24–48 hours for digoxin, and it delivers an increased risk of side effects. Digoxin has one of the narrowest therapeutic ranges of any drug on the market – just 20–50 times a normal dose can prove fatal. Such a small gap between a healthy and a dangerous dose would not normally be tolerated in a new drug coming to market, but digoxin is considered an essential medicine because, with careful monitoring of the patient, the benefits greatly outweigh the risks.

How digitalis kills

Digoxin and digitoxin are completely absorbed through the gastrointestinal tract, so they can be administered in tablet form or as drops as well as by injection. The drugs primarily affect the action of the heart, with injections acting within seconds, but doses administered by mouth may take an hour or so to be absorbed and take effect.

The heart is effectively two pumps working to move blood around the body. The right side of the heart pumps venous, de-oxygenated blood to the lungs, where the red blood cells collect oxygen. This oxygenated blood then moves to the left side of the heart, where it is pumped out to the rest of the

---

*Distinguishing between the different digitalis compounds, either by name or chemical structure, can feel like a game of spot-the-difference. I am still amazed how such tiny differences can result in dramatic increases in potency.

body, to deliver the oxygen necessary for our cells to produce energy. Each side of the heart has two chambers, the atrium and the ventricle. Blood enters the atrium, then moves into its corresponding ventricle. The contraction of the ventricles pumps the blood on to its destination, either the rest of the body (from the left side) or the lungs (from the right).

Digitalis compounds affect the heart in two ways: they intensify its contractions, and they reduce the transmission of electrical signals from the atria to the ventricles that coordinate these movements. Many of the toxic effects of digitalis compounds represent exaggerations of the normal activity of the heart.

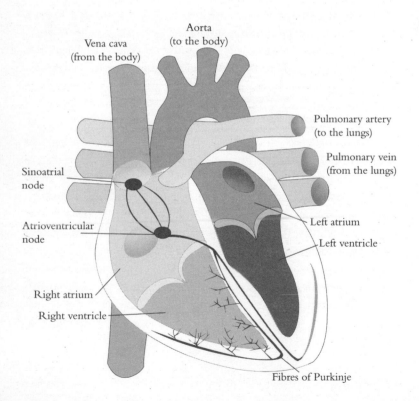

The human heart.

In some cases, the electrical impulses that coordinate the heartbeat across the atria can become disorganised, and irregular pulses are subsequently transmitted to the ventricles. Cardiac glycosides slow the transmission of these electrical signals across the heart; these compounds are therefore of benefit to patients suffering from atrial fibrillation, the rapid and irregular contraction of the atria. This condition is quite common, and can be effectively treated with drugs. In susceptible patients atrial fibrillation can result in the ventricles beating independently of the atria. This 'heart block' is not normally a serious problem in younger subjects, but it can be a significant complication in older patients with other cardiac disorders. High doses of digoxin completely bar transmission of electrical signals between the atria and the ventricles, which can lead to death in minutes. The heart is effectively paralysed.

The structures within the heart that contract to pump the blood are the cardiac muscle cells, or cardiac myocytes. These are the largest cells in the heart, and they make up the bulk of the mass of the organ. Around 50 per cent of each cardiac myocyte is made of microfibrils, an arrangement of thick and thin filaments that slide over each other to cause the physical contraction of the cell. The coordinated contraction of these cells results in the overall movement of the heart, squeezing the atria and ventricles to pump the blood. The sliding action of the microfibrils is as a result of the movement of sodium (Na^+), potassium (K^+) and calcium (Ca^{2+}) ions.* The movement of sodium into a cell triggers a cascade of events that induces calcium ions also to move into the cell. It is the movement of calcium ions that causes the microfibrils to alter their shape and contract the heart.

The complex chain of events that results in a heartbeat must be carefully coordinated. A group of cells in the heart (at the

---

*An ion in this case is an atom that has lost one or more electrons.

sinoatrial node in the right atrium), and modified heart cells known as Purkinje fibres are capable of spontaneous self-excitation (meaning they can generate a nervous signal without instructions from the brain); this triggers an impulse, which travels through the heart in a highly coordinated manner. These cells and fibres in the heart act as a pacemaker; they will continue to function so long as there is a supply of oxygen, plus sodium, potassium and calcium ions and a few other key minerals. Strictly speaking, the rest of the body, including the brain, is not really required for the heart to function. Transplanted hearts will begin to beat as soon as they are connected up to the recipient's blood supply; this independent nature of the heart enables the study of its function in isolation from the rest of the body.

Even in the nineteenth century, frog hearts could be kept beating for a considerable time whilst experiments were carried out directly on them by using a relatively simple solution of potassium and sodium salts. However, it was not until 1880 that the importance of calcium ions was appreciated. This discovery was made accidentally by Sydney Ringer (1835–1910), a physician at University College, London, with some inadvertent help from his laboratory assistant. Ringer made up his regular solution of salts in pure distilled water, to ensure he knew the exact composition of the fluid he was bathing the frog hearts in. One day, a solution prepared by his laboratory assistant caused a frog's heart to beat for many hours, far longer than Ringer had ever achieved in the past. It emerged that the assistant had been rather slack in his preparation of the solutions, and had used ordinary tap water instead of distilled water. The calcium dissolved in the tap water was the key ingredient the heart needed.

One of the effects of cardiac glycosides such as digitalis compounds is to increase the availability of calcium ions in heart cells, thereby increasing the force with which the heart contracts. The effect of cardiac glycosides on calcium ions is actually achieved indirectly by the drugs' interaction with an enzyme, Na^+/K^+-ATPase,[*] which leads to a higher concentration of calcium ions inside the cell, making the microfibrils contract more strongly.

The combined effects of digitalis compounds on the heart cause slower and more intense contractions of the heart – and an increased efficiency in pumping the blood around the body. This improved efficiency increases the rate of fluid being expelled from tissues and increases urine output, hence Withering's observations of the drug's effectiveness in the treatment of dropsy. But digitalis also has significant side effects because the target enzyme, Na^+/K^+-ATPase, is widely distributed throughout the body. The most common side effects are nausea and loss of appetite, though this does not seem to have applied to Mrs Boynton, the victim in *Appointment with Death*, who is described as a large woman. Of course her size may have been due to oedema rather than a healthy appetite. Interactions with Na^+/K^+-ATPase in the brain and the eyes produces side effects such as disturbances to vision, which affect many people taking digitalis compounds, and delirium, which tends to affect older patients.

The concentration of Na^+/K^+-ATPase is particularly high in the cells of the retina in the eye (there is more of the enzyme there than in brain cells, for example). There are two types of cell in the retina that are sensitive to light and are responsible for our vision: these are rod cells and cone cells. Rod cells are responsible for vision in low light. They can detect a single photon, but cannot distinguish between different wavelengths, making the world appear in shades of grey in low light. Cone cells are less sensitive to light, but the three types present in our retinas are sensitive to three different wavelengths

---

[*]$Na+/K+$-ATPase is a bit of a mouthful; it's an enzyme that controls the movement of sodium (Na^+) and potassium (K^+) ions across cell membranes. This movement causes an electrical impulse in nerves, and also triggers the movement of other ions such as calcium (Ca^{2+}) in muscles to cause contractions.

of light,* and are responsible for colour perception. Cone cells are maybe 50 times more sensitive to digoxin than rod cells, so it is colour rather than night-vision impairment that most affects patients. Between 20 per cent and 60 per cent of patients with therapeutic levels of cardiac glycosides in their blood report visual disturbances within two weeks of the onset of treatment, and the most common complaint is colour-vision impairment. Everything appears as though through a yellow film ('xanthopsia'), and there may be hazy or snowy vision. Flickering lights or coloured spots and points of light that appear to have haloes of colour around them are less commonly reported. There can also be effects on the pupil of the eye, such as dilation, constriction or uneven sizes of the pupils, though these are rarer.

Digitalis intoxication may have been responsible for Vincent van Gogh's 'yellow period', as well as for paintings such as *Starry Night*, which some see as exhibiting the 'halo' effect around the stars in the sky. Van Gogh may have been treated with digitalis to control the epileptic fits he suffered towards the end of his life. There is some circumstantial evidence to support this theory; there are two portraits of Paul Gachet, the doctor who treated van Gogh in his last months. In both of these, the doctor is portrayed with foxglove flowers. Also, a self-portrait shows van Gogh's pupils being different sizes; could this be a slip of the brush, or evidence of something more sinister? There is no record of digitalis being prescribed to van Gogh, however, and the foxgloves in the doctor's portraits may be a coincidence. The artist may just have had a strong liking for the colour yellow and the artistic effect of haloes around the stars in the sky, rather than being under the vision-warping influence of digitalis.

Digitalis would have had no effect on van Gogh's epilepsy, but at that time drugs that had proved effective in treating one condition were often prescribed for others, in case they proved

---

*These are 500–700 nanometres (nm) (red), 450–630nm (green), and 400–500nm (blue).

to be a panacea for all ills. There is some scientific credibility to this way of thinking. Many drugs originally developed for a specific treatment have been found subsequently to be more effective in treating other conditions. Viagra is a classic example. This drug was originally developed to treat angina, by relaxing the coronary vessels that supply blood to the heart. The target enzyme was phosphodiesterase; this inactivates a messenger molecule* that causes the dilation of blood vessels, amongst other roles. The drug is effective at increasing blood flow, but it turns out that the heart is not the organ most affected. In addition to its qualities for enhancing men's sex lives, Viagra can affect the dilation of blood vessels throughout the body; those taking heart drugs or who have low blood pressure shouldn't use Viagra as there will still be some interaction with the heart. The drug's effects on blood vessels in the brain can also cause headaches, but its most common side effect is a blue tint to colour perception, caused by its interaction with phosphodiesterase in the cone cells of the eyes.

Is there an antidote?
In an emergency case of digitalis overdose, atropine can be given, to stimulate the heart. In the case of a dose of digitalis that does not kill the patient quickly, potassium chloride (KCl) dissolved in fruit juice can be given every hour until the normal heart rhythm is restored, but the individual must be carefully monitored to avoid potassium toxicity. Complete inactivity is enforced for patients, especially ones with impaired kidney function who cannot process the potassium chloride or other drugs efficiently. The patient is kept completely still until normal heart function is restored.

Today, more drugs are available to treat overdoses of digitalis, and the chances of surviving a huge overdose are much higher. Phenytoin can be given in cases of acute digitalis poisoning, for example, if the patient is unresponsive to potassium therapy.

*Known as cGMP – short for cyclic guanosine monophosphate.

This drug increases the metabolism of the digitoxin in the body, rendering it ineffective more rapidly than under normal circumstances. Another drug, cholestyramine, reduces the half-life of digitalis drugs in the body. Specific digoxin antibodies are also now available to inactivate the drug in the body.

Some real-life cases

Digitalis-based murder cases are extremely rare – or perhaps, as I mentioned earlier, the murderers have chosen their victims carefully, and the deaths have not been treated as suspicious. Agatha Christie had just one recent case to use as a possible inspiration; it occurred two years before the publication of *Appointment with Death*.

In 1932, Marie Alexandrine Becker was a 55-year-old housewife who had decided that her life needed a little more excitement. Her respectable but dull life in Liège, Belgium, changed one day at the market, when she met Lambert Beyer at the vegetable stall. Beyer was something of a local lothario, and the two embarked on a tempestuous affair. The affair changed Marie from a respectable housewife to a serial killer. Her first victim was her husband, for whom she obtained a substantial payment from a life-insurance policy. Soon Marie tired of her new lover, or perhaps the money he had willed to her was too tempting; in any case, Beyer himself became her second victim.

With the money she had obtained, Marie bought a dress shop to fund her new and extravagant lifestyle. But the income was not enough to pay for the nights out in clubs or the payments to the young men that shared her bed. When one of Marie's friends became ill with a dizzy spell she offered to nurse her. Perhaps unsurprisingly, her friend's health deteriorated, and she died a few weeks later. Marie continued to poison friends to obtain money; when she ran out of friends, she turned to the customers in her shop. Apparently she would drop digitalis into cups of tea in a back room and offer them to the ladies picking out dresses in the shop. After they died, she would take any ready cash and valuables that they had on them.

Rumours had already started about Marie, and anonymous letters had even been sent to the police suggesting that she might have been involved in the deaths of two elderly women. An investigation was started, but it was evidence brought forward by a (surviving) female friend of Marie's that motivated the police to look deeper into the case. The friend had been complaining to Marie about her husband, and how she wished the no-good rascal dead. Marie helpfully offered her friend a powder that would dispatch the man, leaving no trace. After thinking over her options for a few days, the woman went to the police.

Marie was arrested and the bodies of her husband, her lover, her friends and her customers were exhumed. Traces of digitalis were found in the bodies, though no suspicion of foul play had arisen at the time of their deaths. Marie was put on trial in 1936 for the murder of ten people, though it is believed she may have killed twice as many. The jury returned a verdict of guilty and she was sentenced to life imprisonment, there being no death penalty in Belgium at the time. She died in prison during the Second World War. Apparently she revelled in the details of her victims' demise, describing one victim as 'dying beautifully, lying flat on her back'.

Dame Agatha and digitalis

In *Appointment with Death* Mrs Boynton and her family take a trip to the deserted city of Petra. Mrs Boynton is a monster of a woman; there is no shortage of people who would be happy to see her dead. Mrs Boynton also has a heart condition for which she is taking digitalis medication, in the form of a solution added to water. Her medical condition and treatment provide an excellent opportunity for her killer to cover their tracks. An overdose of digitalis results in failure of the heart to contract properly, ultimately leading to cardiac arrest. Dr Gerard, who is also on the expedition to Petra, correctly points out that 'A large dose of digitoxin thrown suddenly on the circulation by intravenous injection would cause sudden death by quick palsy of the heart. It has been estimated that 4mg might prove fatal to an adult man.'

There was a plentiful supply of digitalis, despite the fact that
the death occurred in an isolated part of the world surrounded
by desert, where foxgloves would not grow. The killer had no
need to find a local pharmacy and fake a prescription, nor to
search for foxglove plants with which they could make their
own lethal preparations. The poison could have been obtained
from Mrs Boynton's own medication, or from a supply of
digitoxin in Dr Gerard's medical bag.

Mrs Boynton's death could have been due to natural causes,
or accidental poisoning. When her body is found, Dr Gerard
first assumes that her death was due to the exhausting journey
to Petra, combined with the hot weather, with this all being
too much for her heart. Her death could also have been due to
a mistake made by the dispenser when making up her
prescription of digitalis medicine (Christie of course knew
such blunders were possible from her dealings with Mr P.; see
page 11). In Mrs Boynton's case we will never know if the
dispenser made an error, as the bottle containing her medication
was inadvertently broken when her body was moved.

Mrs Boynton dies while the rest of the party are away from
the camp, so there are no witnesses to her final moments.
Doubts about a natural cause of death are raised when
Dr Gerard notices a mark on Mrs Boynton's wrist, the sort of
mark that would be left by a hypodermic syringe. He also
notices that the bottle of digitoxin in his medicine bag has
been noticeably depleted, even though he hasn't used any
during his trip, and his hypodermic syringe is missing. To
clarify the cause of death requires a post-mortem examination
of the body as a first step, but conducting a post-mortem in an
ancient abandoned city is not practical, so Mrs Boynton's body
is taken to Amman … which happens to be the city in which
Hercule Poirot is on vacation.

As Agatha Christie correctly points out, 'the active principles
of digitalis may destroy life and leave no appreciative sign'.
Even so, a large dose would be detectable if the pathologist
knew to look for it. The presence of digitalis was first detected
in a murder victim in 1863; scientific evidence helped convict

Dr Edmond–Désiré Couty de la Pommerais of the murder of his former mistress, Madame Séraphine de Pauw. Pommerais had convinced Pauw to take part in an elaborate insurance swindle in order to pay off his debts. A number of large insurance policies were then taken out on Pauw's life. Pommerais told her that his plan was to convince the insurance companies that Pauw had a terminal illness and would die soon, at which point they would claim a large annuity until she died rather than pay out the vast sums of insurance money. She would then miraculously recover to live the rest of her natural life in financial security. Séraphine even told her sister of the ingenious plan; but the sister saw through Pommerais's promises, and warned Séraphine that he might be planning to kill her and keep all the money for himself.

This is exactly what happened. On 16 November 1863, Pommerais gave Pauw something that made her very ill but, as was predicted by her sister, Madame Séraphine did not recover. Pommerais filed his claims with the insurance company and sat back, presumably reassured that the poison he had chosen to kill his victim could not be traced. The police, however, were suspicious of Pommerais's behaviour; they asked Ambroise Tardieu (1818–1879), a respected medical doctor, to analyse Pauw's body for signs of poisons. After eliminating metals such as arsenic and lead, Tardieu turned to the alkaloids. Using the Stas method (see page 61), Tardieu managed to extract a bitter-tasting substance from Pauw's remains. However, Tardieu could not identify the substance; it was not an alkaloid with which he was familiar. After a series of fruitless experiments, and almost at his wits' end, Tardieu decided to inject five grains of the extract (approximately 300mg) into 'a large vigorous dog' to see what would happen. The answer was – absolutely nothing, for two and a half hours. Then the dog suddenly vomited and lay down, obviously weak. The dog's heart slowed, beat irregularly and occasionally stopped until twelve hours later, when the beast began to recover. Looking at correspondence between Séraphine and Pommerais, Tardieu found discussions of a prescription of digitalis she was taking to 'stimulate herself'. This had all been part of the ruse to obtain money from the insurance companies, but it gave Tardicu the clue he needed – the victim had died from digitalis poisoning.

Tardieu had not recovered enough digitalis from Pauw to account for her death. He explained to the police inspector that what he really needed was a sample of her vomit; this would contain a much higher concentration of the poison, enough perhaps for him to establish a cause of death. The inspector responded to this in a remarkable way. No samples of vomit had been retained, so he went back to Pauw's bedroom, and removed floorboards and wood shavings from parts of the floor where vomit had been spilled. Tardieu quickly set about analysing the samples sent to him, and obtained far greater quantities of the poison from the vomit that had dried onto the floor. To prove the poison was the same as that found in the body, Tardieu observed the effects of the extracts from the floorboards on frog hearts, where he witnessed the same reduced heartbeat. He carefully repeated his experiments, and requested more samples of floorboards from under the bed, where no vomit could have reached – Tardieu wanted to ensure that the effect on the frog hearts was being caused by the poison, and not by varnish or paint from the floor. At the subsequent trial Pommerais's defence team attempted to discredit the scientific evidence Tardieu presented, but it was compelling; Pommerais was executed for his crimes.

By the time Agatha Christie was writing *Appointment with Death*, a chemical colour reaction had been developed to detect the presence of digitalis glycosides. Unfortunately the reaction, which produced a characteristic indigo-blue or bluish-green colour, only worked in the presence of large quantities of the drug, far greater than the amounts needed to kill. Digitalis also decomposes in the body after death, so the best way to confirm its presence was by testing vomit from the victim. If vomit was unavailable then physiological experiments on frog hearts remained the best method for detecting small quantities of digitalis extracted from a corpse. Agatha Christie makes no mention of Mrs Boynton having vomited, and it seems unlikely that she did, as her dead body is mistaken for a sleeping one. Had Mrs Boynton been sick it is reasonable to assume that someone would have called a doctor far sooner than they did.

Today, methods of detection have improved and the means of identifying compounds within the body are standardised; we no longer rely on experiments on frogs. Analysis of the blood is the standard procedure, and even the minute quantities prescribed for daily medication are detectable. Digoxin levels of 0.6–2.6ng/ml* are considered to be at a therapeutic level, and above 2ng/ml would be considered toxic, though individual responses to the drug vary greatly. The drug is released from the muscles of the heart after death, so levels may be unnaturally elevated above those expected for therapeutic doses. In the case of Mrs Boynton, it would have been possible to prove the cause of death using 1938 scientific methods; a post-mortem is ordered, but Poirot doesn't bother to wait for the results. His superior brain is capable of solving the crime before it is confirmed that a crime has even been committed.

The obvious suspects in the case are Mrs Boynton's family. Her daughter-in-law is responsible for measuring out the digitalis mixture that Mrs Boynton took daily, though the exact composition of the prescription is not mentioned. She may have been taking digitalis, a mixture of all the cardiac glycosides extracted from foxgloves, or she may have been taking a purified form of just one of them. Any one of the family could have tampered with her medicine and increased the dose by concentrating the drug, resulting in a higher than normal dose being given. Alternatively, they could have diluted the drug or substituted a placebo, causing Mrs Boynton's heart condition to worsen in its absence. Digitalis can act as a cumulative poison because it has such a long half-life in the

*ng = nanogram. A tiny amount, in other words.

body, but unless the dose was significantly larger than usual Mrs Boynton would have been expected to show a gradual increase in the symptoms of digitalis intoxication, such as problems with eyesight and an irregular heartbeat, rather than a sudden collapse.

Poirot has his doubts about someone having tampered with the medication, not only because Mrs Boynton's death was very sudden, but also because of the mark of the hypodermic syringe on her wrist. He finds it implausible that a member of the family would go to the trouble of stealing a syringe and a supply of digitalis when they had far easier access to Mrs Boynton's own medication. The fact that the fatal dose was administered by hypodermic syringe suggests that another member of the party at Petra was the murderer. What Poirot does not find implausible is that someone who wanted to kill Mrs Boynton, and happened to bump into her on holiday, knew her well enough to know about her health as well as the effects of an overdose of digitalis. The murderer was also fortunate to be on a trip with a doctor who had a supply of the drug in his bag, and was prepared to risk being seen stealing the drug and the syringe from the doctor before finding an opportunity to administer the poison unobserved.

The circumstances that lead to the death of Mrs Boynton may be slightly implausible – in the stage adaptation of *Appointment with Death* Christie changed both the motive and the murderer. But her scientific descriptions throughout the book are top-notch.

Crooked House

If Brenda were to make a mistake and inject eyedrops into me one day instead of insulin — I suspect I should give a big gasp, and go rather blue in the face and then die, because you see, my heart isn't very strong.

Agatha Christie, *Crooked House*

In the quote above, Aristide Leonides is describing the details of his own death. He dies when he is injected with eserine eyedrops, which had been substituted for his regular dose of insulin. Eserine is an unusual choice of poison; Agatha Christie used it in only two of her novels, *Crooked House* and *Curtain*. You might expect a poisoner to have a detailed medical or pharmaceutical knowledge in order to choose eserine, but in *Crooked House*, Christie ensures there are plenty of suspects by giving Aristide a big audience when he speaks these prophetic words. Aristide is actually answering a question from his granddaughter as she looks at his bottles of medicine – 'Why does it say "Eyedrops – not to be taken" on the bottle?' His extended family seated around him all hear his reply; one of them goes on to use the old man's own medicine to kill him.

The list of suspects in *Crooked House* is a long one: there is Aristide's second wife Brenda; his *first* wife's sister, Edith; his sons, Philip and Roger; his sons' wives, Magda and Clemency; and his three grandchildren, Josephine, Eustace and Sophia. There is also the household staff, including the cook, the tutor and the nanny; they all live together with the 'crooked man' Aristide in his 'crooked house'. But this time there is no Hercule Poirot or Miss Marple on hand to help identify the culprit. Instead it is down to the detective powers of the police and Sophia's fiancé, Charles Hayward, to do their best. In the novel, Christie scrupulously presents every clue to her readers, and even hints strongly at the murderer. However, when the culprit is finally revealed it still comes as a shocking revelation (Christie's publishers even had the nerve to ask her to change it, but she refused).

The poison used, eserine, had come a long way to arrive in the Leonides household. This compound is an extract from the beans of a West African plant, *Physostigma venenosum*. Eserine has had a variety of applications in medicine, from the management of glaucoma to emergency treatment for nerve-gas poisoning. There have been cases of accidental eserine poisoning, but criminal poisonings are exceptionally rare, perhaps owing to a combination of eserine's low profile and because treatment is readily available and very effective; eserine victims have a good chance of survival.

The eserine story

Physostigma venenosum is a climbing perennial plant that produces large fruits, each of which contains two or three beans. The use of this bean as a poison was once common in the south-eastern part of Nigeria known as Calabar, and the beans are commonly known as Calabar beans. When the active compound within the Calabar bean was isolated it was named physostigmine. The local name for the bean was *eséré*, so the same compound is also known as eserine.*

The plant, and the poisonous properties of its beans, was well known in West Africa, but it was only brought to the attention of the wider world by Scottish missionaries who arrived in Calabar in the 1840s. Rather melodramatically, the missionaries named it the Ordeal Bean of Old Calabar. They discovered a tradition of trial by ordeal for crimes such as witchcraft, murder and rape. Guilt or innocence was determined by the accused drinking a concoction of poisonous beans. If they were guilty, the beans would poison them and they would die, but if they were innocent they would be spared death. However, anyone who showed ill effects but survived was subsequently sold into slavery. The method of preparing the poison varied. Sometimes the whole bean was used, while at other times the beans would be mashed in water, and the accused would be required to drink the resulting milky liquid. These trials by bean were known locally as 'chop nut'.

There may be more to this trial by poisonous bean than initially meets the eye. Guilty people may have been more inclined to chew the beans slowly and prolong the moment before they had to swallow. Chewing would release more of the poison than if the bean was swallowed whole. Or, if a drink was given, the guilty might sip it cautiously. The end result was the same; slow administration of the poison and a longer exposure time, giving the body more opportunity to absorb the poison. Innocent people would swallow their beans quickly, safe in the knowledge that they were not guilty. The whole bean would take longer to digest, and less of the poison would be released. Alternatively, the beans could cause gastric irritation, inducing vomiting that saw the poison largely removed before it could be absorbed.

*The name most commonly used today is physostigmine, but eserine was the name Christie used in *Crooked House*, so for convenience I'll stick with that for the rest of this chapter.

Other theories have been suggested that throw a different light on the trials. Those responsible for acquiring the beans and preparing them would be expected to have a greater knowledge of native plants and their effects on humans than the general populace. Half a Calabar bean can constitute a lethal dose, but the quantity of poison it contains varies depending on ripeness. Beans could be harvested at different times to affect the result of the trials, and the beans themselves could be doctored to give the desired result. The small brown beans could also have resembled non-toxic varieties, and a substitution could have been made that might have gone unnoticed by the accused.

Local people were so confident in chop nut that many actively volunteered to swallow the poisonous concoctions to prove their innocence. The missionaries claimed that thousands gathered at witch trials to swallow beans *en masse*. The beans were also used in duels as well as in trials. Combatants would cut a bean in two, and each eat a half. They would continue to eat the divided beans until one or other (though quite often both) died as a result. The missionaries estimated that the Calabar bean was responsible for 120 deaths in the region every year.

The missionaries were keen to collect samples of the bean, as well as of the whole plant, to send back home. Collectors travelling around the world had brought exotic delights such as the coffee bean and cinchona bark (which contains the active component, quinine, used in the treatment of malaria) to Europe. These discoveries had given us new foods and medicines, in addition to a range of plants to adorn gardens. Many people built careers out of collecting exotic specimens and selling them to buyers in Europe, but the missionaries in Nigeria were having problems tracking down samples of the *eséré* bean and the plant it came from. Initially they could only obtain a few leaves. Eventually the Reverend Hope Waddell, a missionary who had been apprenticed to a pharmacist before he became ordained, discovered that the king in Calabar had ordered the wholesale destruction of the plants except for a few specimens kept under close guard, for use in administering local justice. In 1855 Waddell smuggled a few beans out of the country and sent them to Robert Christison (1797–1882), a celebrated Edinburgh toxicologist. Christison managed to grow a few plants from the beans, but none would flower. It wasn't until 1859 that flowering specimens were brought from Nigeria for study; the king in Calabar had either relaxed his bean embargo or some unscrupulous soul had found a way around his strict controls.

Christison was interested in the medical effects of the beans, and he took the then traditional method of testing toxins by swallowing a quarter of one. The beans, Christison noted, had a very bland taste and he initially suspected they were harmless. In his own words he was 'much mistaken'. The most dramatic effect Christison experienced was his heart slowing, and he suggested that death may be caused by paralysis of the heart. Some years later, in 1897, John Uri Lloyd (1849–1936), an American pharmacist, suggested that the bean could be used to execute condemned criminals, as the poison appeared to act painlessly. The suggestion was not taken up.

In 1863, the ophthalmologist Douglas Argyll Robertson (1837–1909) published a paper describing the effects of an extract of the bean on the pupils of the eye, and this spurred further research. Eserine was the first known miotic, a compound that causes the pupil to contract. Robertson openly admitted that he had been told about the unusual properties of eserine by a physician friend, Thomas Fraser (1841–1920), who detailed the process of extraction in his doctoral thesis. Even after he had completed this, Fraser continued to carry out rigorous and detailed studies on eserine's properties, and other scientists joined in. By the late nineteenth century many alkaloids had been extracted from Calabar beans, but the structures of these compounds were not elucidated for another 30 years or more. In the meantime, researchers could only judge the contents of their extracts by their physiological effects, and there were still problems with extracting the active compound from the bean and storing it without degradation;

eserine is unstable in water and decomposes to form eseroline. Eseroline has a very different effect on the body, delivering pain relief through its interaction with opioid receptors (see page 184), amongst other diverse effects. Despite the problems faced by Victorian Calabar-bean researchers, many discoveries were made.

Scientific interest in the Calabar bean is reflected in Agatha Christie's novel *Curtain*. One of the characters in the novel, Dr John Franklin, a research chemist, conducts experiments with extracts of the bean. Christie goes into a detailed description of the poison and its uses (but here she calls it physostigmine). Dr Franklin is assisted by Judith, the daughter of Captain Hastings, friend of Hercule Poirot and narrator of *Curtain*. He quotes his daughter talking about the research: 'She referred learnedly to the alkaloids physostigmine, eserine, physovenine, and geneserine, and then proceeded to the most impossible-sounding substance, prostigmin or the methyl carbonic ester of 3-hydroxyphenyl trimethyl ammonium, etc., etc.'

Agatha Christie's science is good here, and there are only minor errors in her list of alkaloids. The confusion is understandable, though, as many of these compounds had undergone numerous name changes over time. As we've seen, physostigmine and eserine are the same compound, but Christie might have been confusing eserine with eseramine, which is also present in the bean. Another compound she lists, prostigmin (now known as neostigmine), is a synthetic derivative of eserine. Neostigmine was first prepared in 1931, in an effort to modify eserine into a compound that was more stable in water but that retained the same physiological effects. This work was successful, and the resulting compound was found to be a more effective miotic agent than the parent compound.

Of the other alkaloids Christie mentions, geneserine is the second commonest alkaloid within the Calabar bean (only 35 per cent as much as the quantity of eserine). Geneserine is sometimes prescribed as drops to be taken orally for the treatment of digestive disorders such as dyspepsia and constipation.

Meanwhile physovenine, also mentioned by Christie, has been found to be effective in treating the symptoms of Alzheimer's disease.

In *Curtain*, Dr Franklin claims that there are two types of beans that resemble each other very closely. He claims the second type contains all the same alkaloids as the Calabar bean but with one extra, and that this additional alkaloid neutralises the effects of the others. The existence of a plant similar to *Physostigma venenosum* is hinted at in the correspondence of a Mr William Milne, who was living in Calabar in the 1860s. One species was 'largely cultivated for putting into streams to kill fish, and another is sold in the markets for Calabar chop'. It would make sense if the species used to kill fish was less toxic to humans than *Physostigma venenosum* – people would want to eat the fish, after all.

Dr Franklin goes on to describe how the bean was used in West African ordeal trials, and his belief that those in the inner circle knew of the second type of bean, and used it in a secret ritual. These people never came down with 'Jordanitis', a disease Christie seems to have invented to give a reason for Franklin's research and his possession of many bottles of potentially lethal extracts of the Calabar bean.

To obtain the poison, the murderer in *Curtain* had no need to travel to Africa or to fake a prescription; all they had to do was take a bottle of an extract from Franklin's laboratory. The contents of the bottle are added to an after-dinner cup of coffee. Several hours pass before the victim – the doctor's wife, Barbara – starts to exhibit the effects of the poison, and she dies the following morning. There are no details of the symptoms experienced by Mrs Franklin other than the fact that she is taken 'violently ill'. Little is done to treat Mrs Franklin, although there is plenty of time to call a doctor; a doctor would be likely to have the antidote to eserine poisoning available, in the form of atropine (which we first encountered on page 49). The use of atropine as an antidote for eserine, and *vice versa*, was first proposed by Thomas Fraser back in the 1870s. The effect of contracting the pupil and slowing the heart were directly opposing to the effects of atropine, and each compound could therefore be used to counteract the other's actions. Some doubt was thrown on Fraser's theory because he carried out his atropine/eserine experiments on rabbits, the only mammal he could obtain easily in the quantities he needed; unfortunately, rabbits have a particularly high tolerance of atropine.

However, atropine antidotes were successfully used in several cases of eserine poisoning, and *vice versa*. Two cases of mass Calabar bean poisoning occurred in Britain, the first of these in 1864 and the second in 1871. Despite the best efforts of the king in Calabar, some beans were smuggled out of the country and brought to Britain on ships. Some of the beans were dropped at Liverpool docks during the unloading process, where they were found by some children, who ate them. A total of 57 children became ill, but all but one were saved through prompt treatment with atropine. Even if atropine is not immediately available, the patient can be supported with artificial respiration. Mrs Franklin seems to have been particularly unfortunate.

The main symptoms exhibited by the children at the Liverpool docks, and presumably by Mrs Franklin in *Curtain*, were tremors, involuntary defecation and urination, pinpoint pupils, difficulty in breathing and a slow pulse. The Liverpool children were apparently quite docile and far from distressed by their situation, and they did not cry or seem unduly agitated during their time in hospital. This supports earlier descriptions of the painless effects of this poison. Individually, the poisoning symptoms can be attributed to a wide variety of diseases and toxins, but this combination is characteristic of eserine and related compounds.

How exactly eserine produced these effects on the body was unknown to Victorian scientists, but there was a lot of speculation. For example, the tremors were clearly the result of interference with the nerves, but was this a result of direct action on the spinal cord or a secondary effect as a consequence of interaction with nerve endings? These scientists were working at a time when the mechanisms of nerve function were still a mystery, so it is hardly surprising that they struggled to explain how eserine interacted with the body. When the mechanism of nervous signalling was determined in the 1920s, eserine provided scientists with vital clues to solving the mystery.

Before 1921, there were two schools of thought about how signals were transmitted between nerves, and between nerves and muscles. At these connections there is a gap called the synapse. It was known that electrical signals travelled along nerves, but could electrical impulses also be responsible for signals crossing the synapse? If it was not an electrical impulse, could it be chemical signals traversing the gap? The experiment that would decide which was correct came to scientist Otto Loewi (1873–1961) in a dream. He hastily jotted down the idea in the middle of the night and went back to sleep, but the next morning he found he could not read his writing, and couldn't remember his dream. The next day was the longest of his life as he tried to recall his moment of inspiration. Fortunately the following night *he had the same dream*. This time he got up and went straight to the lab.

Loewi dissected out two frogs' hearts and placed them, still beating, in Ringer's solution (see page 98) in separate dishes. To the first heart he applied an electrical current that slowed the heart. He then took the fluid surrounding the first heart, and transferred it to the dish containing the second. The second heart slowed too. The electric stimulation of nerves in the first heart had triggered the release of chemicals that could affect the action of nerves in another heart. Transmission of signals across the synapse therefore had to be as a result of chemicals. Now that the general method of getting signals across synapses was understood, the details would be much easier to work out.

The most pressing issue was to establish which chemical was being released by the nerves. Loewi and his team knew two things: that whatever it was, it disappeared very quickly; and that its effects were blocked by atropine. After testing various substances known to act on the nerves, including muscarine, pilocarpine, choline and acetylcholine, they found that acetylcholine matched all the criteria – swift disappearance from the body, and blockable by atropine. Further experiments, published in 1926, revealed that the reason acetylcholine seemed to disappear was that it was being broken down by an enzyme, cholinesterase. Loewi and his collaborator E. Navratil found that eserine inhibited this enzyme, stopping the breakdown of acetylcholine and allowing it to continue to interact with receptors. The use of eserine allowed scientists to study the effects of acetylcholine more closely, before it was broken down and disappeared.

The use of eserine had provided valuable information about the mechanism of nerve-signal transmission. This work was recognised by the Nobel Committee in 1936 when it awarded Loewi the Nobel Prize for Medicine and Physiology. As Loewi pointed out in his Nobel Prize lecture, the operational mechanism of an alkaloid had been determined for the first time. The initial work on eserine led to our understanding of how many other compounds, such as the organophosphates used as nerve gases and insecticides, inhibit cholinesterase in humans and insects (see page 65).

Acetylcholine – one of a number of chemicals known as neurotransmitters – is released predominantly by nerves in the parasympathetic nervous system or PN, the 'rest and digest' system that regulates fluids such as tears and saliva (though it is also found in both the sympathetic and central nervous systems). Eserine therefore mainly affects the PN. Another of the PN's roles is stimulating contractions of smooth muscle – in the gastrointestinal tract (allowing food to be squeezed through the gut), in the urinary tract and in the eye, as well as decreasing heart rate and relaxing the smooth muscles of blood vessels.

The breakdown of acetylcholine after it has performed its task is vital. To stop receptors from being constantly activated, and to allow repeat signals to be sent to the same site, the acetylcholine must be removed from the binding site so the receptor is ready to receive more signals. The body uses enzymes, cholinesterases, to break up the acetylcholine, using water to cleave chemical bonds and leave two compounds, acetate and choline, neither of which will interact with the receptor site; in other words, going back to our 'lock and key' analogy (see page 37), the key is removed from the lock, leaving it ready for the next one to enter.

The human body has two types of cholinesterase enzymes: acetylcholinesterase (AChE) and butyrylcholinesterase (BChE). AChE acts almost exclusively on acetylcholine, and is predominantly found in the muscles and brain. The body contains more of the BChE enzyme, though, which is distributed throughout the body. The BChE lock can be opened by several different molecular keys, whereas the AChE lock is specific to the acetylcholine one. BChE acts on a range of compounds, including aspirin, cocaine and heroin, breaking them down and limiting the amount of cholinergic[*] toxins that can reach the brain.

Eserine binds to AChE as if it were acetylcholine, but the structural differences between the two compounds means a different chemical reaction then takes place. The eserine is broken up by the enzyme but only slowly, and in the process a fragment of the eserine structure known as a carbamate unit is transferred to the active site of the enzyme. With the carbamate present, the enzyme cannot carry out its normal function and is effectively inactive. It is as though a key has broken in the lock and a fragment of it has got stuck, so that no other keys will fit in the lock until the fragment is removed. The enzyme can be regenerated by having the carbamate removed by another enzyme, but this process is also slow. While the AChE is out of action, acetylcholine will continue to interact with the nerve receptors, and will stimulate them.

Eserine is classified as a 'reversible cholinesterase inhibitor' because the enzyme can regain its function. There is an 81 per cent

[*]'Cholinergic' means anything that mimics the action of acetylcholine.

recovery of AChE after two hours and 100 per cent within 24 hours. By contrast, other AChE-inhibiting compounds, such as the organophosphate nerve gas sarin, bind permanently; eserine can therefore prevent sarin poisoning, if it is administered at the appropriate time, by temporarily blocking AChE until the body has had a chance to eliminate most of the sarin. Eserine is also more soluble in fats than many other AChE inhibitors, so is able to cross the blood–brain barrier to prevent damage to the brain in the event of poisoning by sarin or similar compounds.

As well as its use in cases of atropine and sarin poisoning, eserine was once suggested as a treatment for tetanus,[*] and as an antidote for strychnine and curare poisoning. The most successful of these treatments was with curare and the drugs derived from it. Curare is the plant-derived arrow poison Agatha Christie's pharmacist Mr P. carried around in his pocket (see page 12), and along with its related compounds, it has found wide uses in medicine. These compounds cause the relaxation of muscles by blocking acetylcholine receptors, which is often very useful during surgical procedures.

The similarity between curare poisoning and an inherited condition, myasthenia gravis, led the physician Mary Broadfoot Walker (1888–1974) to test eserine on one of her patients. Myasthenia gravis produces fluctuating muscle weakness, with muscles becoming weaker with increased activity but improving after periods of rest. Symptoms can develop suddenly, and are intermittent. The muscles that control eye movements, facial expressions, chewing and swallowing are usually the worst affected, but movement of the limbs and the muscles that control breathing can also undergo periods of weakness. When Walker conducted her experiments in 1934, the cause of myasthenia was not known, but one theory held

---

[*]Eserine was successfully used in several tetanus cases before vaccinations were available.

that patients were not producing enough acetylcholine to act on the receptors in the muscles. Injections of eserine produced a dramatic, though temporary, improvement in two patients, indicating that although the individuals were producing acetylcholine, it was failing to act on the muscles. Further studies have shown that myasthenia gravis patients produce antibodies circulating in the body that block acetylcholine receptors. Today the condition is treated by a combination of immunosuppressant drugs and anticholinesterases (such as neostigmine) that prevent the breakdown of acetylcholine and allow it to act on the receptors for a longer period of time.

Some real-life cases
Since the missionaries put a stop to ordeal trials in West Africa, deliberate eserine poisonings have been rare outside of Agatha Christie's novels, which makes you wonder where she found her inspiration. The poison itself is difficult to obtain, and even if someone had a prescription for it, consuming a whole bottle would be unlikely to be fatal, though it might make them very ill. In 1968 a biochemistry student attempted suicide by swallowing 1g of eserine salicylate, which he had stolen from a laboratory. He developed severe abdominal pains after ten minutes, followed by terrifying hallucinations that induced him to seek help. Although atropine was administered it made his condition worse, as he did not display the characteristic slow heartbeat, and atropine only increased his heart rate. Atropine is not a *true* antidote for eserine poisoning because the compounds act on different sites in the body. He was subsequently given aldoximes, which reactivated the AChE enzymes that were inhibited by the eserine in his body. He went on to make a full recovery.

I have only managed to find one other case of deliberate eserine poisoning, but it is not clear who did the actual deed. The case was reported in Austria, long after the publication of *Crooked House* and *Curtain*. A man of around 50 years of age was transferred to hospital suffering from diarrhoea and vomiting. He was released after a week of successful treatment.

A month later he was readmitted with the same symptoms. Toxicological analysis of the stomach content detected eserine, and calculations based on 450ml of stomach contents suggested that he had ingested approximately 100mg.

After two months in hospital the patient's condition changed dramatically for the worse, and medical staff were unable to save him. The cause of death was given as cardiogenic shock, but the deterioration in his body brought about by the eserine was thought to be the underlying reason for his death. An inquiry subsequently looked into the origin of the eserine the patient had ingested. Re-analysis of his stomach contents revealed eserine to be the only alkaloid present in the body in detectable quantities. Had the patient been poisoned with Calabar beans, other alkaloids would be expected to have been present.[*]

Pharmaceutical preparations of eserine, sold under the name 'anticholium', also contain very small but detectable quantities of geneserine (the second most common alkaloid in Calabar beans). Anticholium is used to treat atropine poisoning, and is available in 5ml ampoules. The patient would have had to drink 50 ampoules of anticholium to reach the levels of eserine found in his stomach. This seems unlikely; together with the fact that no other compounds (such as geneserine) were found in the stomach, it was concluded that he had been poisoned with the pure chemical.

Agatha and eserine
The victim in the 1949 novel *Crooked House*, Aristide Leonides, is 85. One day, after his usual injection of insulin, he has a sudden seizure. The family can do nothing for him and call a doctor, but by the time he arrives Aristide is dead. He was not a well man, suffering from diabetes, a weak heart and glaucoma, but his death is very sudden and quite unexpected. The

---

[*]Methanol extractions made from beans contain eserine, geneserine (at a quantity of 35 per cent that of eserine) and another alkaloid, norphysostigmine (at a quantity of 12 per cent that of eserine).

symptoms Agatha Christie mentions in the novel are 'difficulty in breathing' and 'a sudden seizure', but in reality there were likely to have been others. Agatha was far too discreet to detail any involuntary urination or defecation, but other symptoms such as a slow pulse and convulsions might have been mentioned without embarrassment to the reader. Anyway, the symptoms are enough to make the doctor suspicious and demand a post-mortem.

The police investigating Aristide's death admitted that little was known about the post-mortem appearances of eserine. There may be congestion in the brain, lungs and gastrointestinal tract, although these signs might be found in cases of poisoning by a range of substances. When combined with the symptoms displayed by Aristide just before death the finger of suspicion would point towards one of the cholinesterase inhibitors, but there are many of those besides eserine. However, the fact that Aristide had a cholinesterase inhibitor in the form of eserine in his medicine cabinet – and an empty bottle of eserine eyedrops was found in the rubbish bins of Aristide's crooked house – gave the pathologist a good indication of which poison to look for at the post-mortem.

Plant alkaloids such as eserine can be extracted from tissue using the method first developed by Stas in 1850 (see page 168). The isolated compound could then be identified by chromatography; in 1949, when the novel was written, chromatographic techniques were beginning to be used more widely. This would enable a toxicologist to see whether a victim had been poisoned by eserine from a medical prescription or by ingesting the bean, as the bean would contain other alkaloids that would also appear in the chromatography. This technique can also determine the amount of a particular compound in a sample. In *Curtain*, analysis of the alkaloids extracted from Mrs Franklin's body showed that several alkaloids of the Calabar bean were present, proving that she had been poisoned with one of the extracts used in her husband's research rather than by prescription medication. In *Crooked House* the post-mortem results confirmed that Aristide had died of eserine poisoning from his medication, as there was an absence of other Calabar bean alkaloids.

Aristide had been taking eserine in the form of eyedrops to treat glaucoma. This is a group of eye conditions that cause the pressure in the eye to build to abnormally high levels. This compresses the optic nerve and can cause permanent nerve damage, resulting in loss of vision. Various treatments are available depending on the cause of the glaucoma, but in the case of acute glaucoma miotic drugs are used to contract the pupils.

There are two sets of muscles in the iris that control the size of the pupil. Radial muscles spread out from the pupils like spokes on a bicycle; their contraction causes the pupil to dilate. Circular muscles that form rings around the pupil contract to make the pupil smaller. The action of the radial muscles is controlled by the sympathetic nervous system, and the circular muscles by the parasympathetic nervous system (PN). As we've seen, eserine acts predominantly on the PN, and therefore causes the iris to stretch and the pupil to contract. By stretching the iris, it is pulled away from drainage channels in the eye, allowing fluid to drain out. Side effects of changes to pupil size include alterations to vision, problems with night vision and problems with focusing. These side effects can be minimised by using appropriate doses, and any inconvenience must be better than the potential loss of sight. Of course there could be other more dramatic and serious side effects caused by eserine's interaction with AChE in other parts of the body, such as the heart and muscles, but these are rare when using small doses applied directly to the eye.

In *Crooked House*, it is clear in the post-mortem examination that Aristide had been killed by an injection of eserine eyedrops, but would this have worked? Eserine can kill in relatively low doses if it is administered by injection because the drug is delivered directly into the bloodstream; if the drug is swallowed or absorbed through mucous membranes, the body has an opportunity to digest and break it down before it reaches the AChE enzymes and starts to cause problems.

Eserine eyedrops today are prescribed in solutions of 0.25 per cent w/v* in 15ml bottles (equivalent to 15mg of eserine per bottle). From this bottle a patient puts a drop or two into each eye, between one and three times a day. The LD_{50} (the amount required to kill 50 per cent of animals – in this case mice – in a test group) by injection is 0.6–1.0mg/kg. For a 70kg adult human this translates to 40–70mg, though the minimum lethal dose recorded is a mere 6mg. This means that a murderer would have to inject between three and five bottles of eserine eyedrops to kill a 70kg man – one would expect the victim to notice, and protest. It is true, however, that Aristide was an elderly man with a weakened heart, and he might have been expected to be killed by a dose at the lower end of the scale. To kill by ingestion, a poisoner would have had to increase the dose and administer approximately 14 bottles, or 210ml, to achieve the same end.

The volume of insulin injected into a diabetic patient varies, but it is generally around 1ml. If Aristide's insulin had been substituted with a modern prescription of eserine, he would have been injected with approximately 1mg of eserine, way below the minimum lethal dose; although Aristide's eserine medication may have been more concentrated than would be expected today, or his insulin injections may have been more dilute, requiring larger volumes to be injected. The time taken for Aristide to show symptoms of poisoning is also curious. He was apparently well for half an hour after his injection. This kind of time delay might be expected if he had ingested the poison, but an injection would have acted within a few minutes.

Eserine is certainly an unusual poison to choose for a murder mystery. It's certainly lethal in very small quantities; Agatha Christie's use of this alkaloid in the novel *Curtain* is highly realistic. However, the method she adopted for its use in

---

*w/v is shorthand for 'weight by volume'. In this case, this is equivalent to 2.5g in one litre.

Crooked House would have been unlikely to have killed Aristide, though it would probably have made him very ill. It is a rare example in Christie's work where the details don't quite add up, but this is quibbling, really; *Crooked House* is otherwise one of her best novels.

HEMLOCK

Five Little Pigs

My heart aches, and a drowsy numbness pains
My sense, as though of hemlock I had drunk ...
John Keats, 'Ode to a Nightingale'

Hemlock has been synonymous with poison and witchcraft since ancient times. This lethal plant famously brought about the death of Socrates in 399BC. It has been mentioned in poetry and prose throughout history – it was an ingredient in the witches' cauldron in *Macbeth*, for example, and has even been incorporated into the name of a *Sesame Street* character, the detective Sherlock Hemlock. Hemlock and extracts from it have been used in traditional remedies for millennia, and it was listed in the *British Pharmacopeia* until the early twentieth century. Considering the plant's notoriety it is surprising that, with the exception of Agatha Christie's writings, there have been no intentional hemlock poisonings since the time of

Socrates.[*] However, there have been plenty of accidental cases, owing to the plant being mistaken for one of the edible varieties of the Apiaceae, commonly known as the carrot or parsley family, to which hemlock belongs; the leaves of hemlock have been mistaken for parsley, the roots for parsnips and the seeds for anise. In 1994, hemlock was reported as the third most frequent cause of plant poisoning. Anyone thinking of foraging for wild parsley or wild parsnips should make absolutely sure they have the correct species before they consume them (and if in doubt, don't eat it or feed it to anyone else).

Agatha Christie made use of hemlock in only one of her novels, *Five Little Pigs*,[†] which was written in 1942. The novel concerns the murder of Amyas Crale, a talented but temperamental artist. His body was found sprawled in front of what was to be his last work, a portrait of the beautiful Elsa Greer. He had drunk from a glass of beer that had been poisoned with hemlock; his wife Caroline was found guilty and hanged. Many years later, their daughter Carla asks Hercule Poirot to reinvestigate the case, as she believes her mother was innocent. To find out what really happened, Poirot interviews 'five little pigs', the five suspects present on the day of Amyas's death. The characteristics of hemlock poisoning provide Poirot with vital clues that allow him to solve the case.

The hemlock story

There are actually several species of plant that are collectively referred to as hemlock; these include four *Cicuta* species of water hemlocks, which grow throughout Europe and North America, and the closely related spotted hemlock, *Conium maculatum*. The plants are similar in appearance and are all highly toxic, but they contain very different poisons. These plants are also similar in appearance to less toxic species such as *Aethusa cynapium*, the poison parsley, which contains

---

[*]None that I have been able to find, anyway.
[†]Entitled *Murder in Retrospect* in the United States.

compounds with similar toxic effects to *Conium* if ingested. The compounds in *Aethusa cynapium* are less potent, however, and less likely to cause a fatality, though this is not a recommendation to try it in a salad. All of these plants are part of a family that includes edible species such as carrots, parsnips and parsley.

Socrates was found guilty of corrupting the minds of the youth of Athens, and of impiety. His punishment was to drink from a poisoned cup. In his dialogue *Phaedo*, Plato describes how Socrates was encouraged to walk around until his legs felt heavy, and then to lie down. He was also told not to speak, as talking was apt to 'raise the heat' and interfere with the action of the poison. Those who excited themselves were sometimes obliged to take a second or third dose, as the jailer only prepared as much poison as he deemed sufficient. Socrates appears to have been indignant regarding the requirement to stop talking, and he told the jailer that he should be prepared to give him poison two or three times.

The death proceeded as the jailer predicted. Socrates walked until his legs began to fail. Numbness spread from his feet to his legs and through the rest of his body. He explained to his friends and pupils, who had gathered around him, that when the poison reached his heart he would die. Socrates was conscious and coherent to the very end. He talked to his pupils and requested that they settle his debts after he had died. There was a slight movement, then his eyes became fixed. His friend Crito then closed his eyes and his mouth.

Confusion over the naming of hemlock plants led many to believe that Socrates had been given a concoction made from *Cicuta* plants, which contain cicutoxin, a stimulant of the central nervous system that causes choking and violent convulsions. However, the symptoms of Socrates' poisoning are inconsistent with this; they simply do not tally with the peaceful death described in *Phaedo*, and many doubted Plato's account. Only in the nineteenth century was the matter cleared

up by Scottish pathologist John Hughes Bennett (1812–1875),
after a terrible mistake. In 1845 Duncan Gow, a poor tailor
living in Edinburgh, was brought a parsley sandwich by his
children. Unfortunately the children had picked hemlock
instead of parsley, and Gow was poisoned. Gow's symptoms
were a slow, progressive paralysis followed by death. There was
no choking and no convulsions, with Gow remaining lucid
almost to the end. Bennett performed the post-mortem
examination, and had the plant material identified. Gow, and
Socrates, had been poisoned with *Conium maculatum*, the
spotted hemlock.

Plato had sanitised the description of the symptoms, perhaps
for the benefit of his readers and probably also out of respect
for Socrates. From Plato's description the process appears
almost painless, but this may not be the case. The slight
movement reported at the very end may have been convulsions,
and he would have gasped for his final breaths as he asphyxiated.
Plato also fails to mention any increased salivation and slurring
of words, probably because it was felt to be undignified, but
the poisoner may have added other compounds to the drink to
counteract these effects. One suggestion is that opium would
have been added to the mixture, to relieve any pain and hasten
death. Other plants such as belladonna (see page 57) would
have dried up any secretions, and these properties would
certainly have been known to the ancient Greeks. The exact
composition of the poisoned drink has not been passed on to
us, but we know hemlock was involved.

Hemlock has also been implicated in the death of Alexander
the Great, though many poisons have been suggested as the
cause of his death, including arsenic and strychnine. Alexander
died after eleven days of fever, aphasia (a disturbance of
expression and comprehension of speech) and weakness in his
limbs – classic symptoms of poisoning by *Conium maculatum*.
Further circumstantial evidence comes from Pliny who
claimed that a letter, written by a physician to Alexander,
recommended that he drink wine as an antidote to *Conium
maculatum*. Alexander died in 323BC and it seems unlikely that

we will ever be able to uncover the truth about his death after such a long time.

The name *Conium* comes from the Greek *konas*, meaning 'to spin like a top', one of the symptoms experienced after ingesting the plant; *maculatum* is from the Latin meaning 'spotted' because of the brownish-reddish spots found on the stem of the plant. As well as the name 'spotted hemlock', *Conium maculatum* is known as poison hemlock, the Devil's bread and the Devil's porridge. The plant is native to Europe, where it grows wild beside rivers, in wasteland and on road verges, but it can be found in many other parts of the globe. The species was introduced to the United States as a garden plant, and has been accidentally transported to other countries when its seeds have contaminated grain.

The seeds germinate very easily, and the plant is a noted 'pioneer' species, being one of the first to grow in disturbed soil, and leading the way for colonisation by other plants. Due to its to ease of germination, hemlock is often one of the first plants to appear in spring, so it is of particular significance to farmers. Animals grazing on land where there is little else to eat can become ill from eating hemlock; it is more poisonous to cows than to any other species. Hemlock causes a staggering gait, increased salivation and tears, and respiratory distress, all of which can lead to death. Cows are hardy and can recover from many of these toxic effects, but some of the compounds within spotted hemlock can produce lasting problems because they are known to be teratogenic. Teratogenic compounds cause abnormalities in the unborn foetus. Skeletal malformations can be present in the calves born to cows that have ingested this plant (and similar effects have been seen in sheep, goats and pigs).

Some other animal species appear to be unaffected by hemlock. Quail, for example, are said to eat hemlock seeds with impunity, but it makes their flesh poisonous. There have been several reported cases of indirect hemlock poisoning in humans through eating songbirds that had fed on hemlock seeds. Seventeen such cases occurred in Italy between 1972 and 1990. There were four deaths, three from renal failure and one from prolonged paralysis.

Symptoms of gastritis and loss of coordination begin to appear around 30 minutes after ingestion of hemlock or its toxic alkaloids. The pulse becomes rapid and weak, vision dims and there is a gradual loss of movement, towards paralysis. Consciousness is maintained throughout, until the victim suffocates from respiratory paralysis.

Conium maculatum contains several toxic alkaloids, but coniine is the most studied, and this compound is used as the means of murder in Agatha Christie's *Five Little Pigs*. All of the hemlock alkaloids are classed as piperidines (because they are all based on the structure of a chemical called piperidine – see Appendix 2). Seven spotted hemlock alkaloids have been identified and isolated so far;[*] the toxicity of hemlock is due to the sum of all the toxic compounds present and their relative abundance can vary hugely. To produce these alkaloids, the plant carries out a series of reactions that subtly modify the structure from one to the next, resulting in a cascade of chemicals. γ-coniceine is the first to be produced, and it is this from which all the other alkaloids are manufactured. Experiments with mice have shown γ-coniceine to be the most toxic of these compounds, and this explains hemlock's particularly staggering toxicity in early spring, when γ-coniceine levels are at their highest. However, the concentrations and relative proportions of the different *Conium* alkaloids appear to depend on factors such as temperature, moisture, time and age of the plant. According to research γ-coniceine is predominant in the rainy season and coniine in the dry, but huge changes in concentrations occur during the

---

[*]These being γ-coniceine, coniine, N-methyl-coniine, conhydrine, psuedoconhydrine, conhydrinone and N-methyl-psuedoconhydrinone.

life of the plant, especially during flowering and fruit-forming stages.

Coniine was the first of the alkaloids to be isolated from hemlock, and it is highly toxic. A dose of 100–130mg of coniine would be fatal for an adult human. Pure coniine is a colourless, oily substance, with a pungent odour sometimes described as being like mouse urine. The oil is quite volatile, and is partly responsible for the characteristic (and unpleasant) smell produced when the leaves of *Conium maculatum* are crushed.

The credit for the discovery of coniine goes to a chemist, L. Gieseke, in 1827, but he did not offer any suggestions as to what the chemical formula of coniine might be. The structure was not confirmed until 1881 by August Wilhelm von Hofmann (1818–1892). Today, there is an array of techniques that can be applied to compounds to identify and determine their exact composition, and the arrangement of atoms within the molecule. In Hofmann's time he would have had to use a series of painstaking chemical reactions, and analysis of the products of those reactions, to identify fragments of the original. Those fragments would then be slowly pieced together, as a detective might string together clues to identify a culprit. This was not easy, even for a relatively simple molecule such as coniine. Five years after Hofmann determined the structure, fellow scientist Albert Ladenburg (1842–1911) devised a method for coniine's chemical synthesis. It was the first plant alkaloid to be fully characterised and synthesised, more than 50 years after it was first isolated.

How hemlock kills
Coniine and the other alkaloids that occur in spotted hemlock are neurotoxins. The toxic effects of coniine are due to the molecule's similarity to nicotine, and it acts in a similar way.

Coniine interacts with receptors in the synapse (the junctions between neurons), those that normally bind acetylcholine (see page 56); specifically, nicotinic-type receptors. The interaction blocks acetylcholine from binding to the receptors and stops signals from being received from the connecting nerve.

The peripheral nervous system (PNS) connects the body to the central nervous system (the brain and spinal cord – the CNS), and acts as a relay for messages to and from the brain and body. One branch of the PNS, the autonomic nervous system, controls automatic functions – heart rate and the production of tears, for example; the other branch, the somatic nervous system, relays messages from the brain and spine to the muscles that control our movement, and is largely under our conscious control. Coniine interactions with receptors in the autonomic nervous system lead to symptoms such as increased salivation, pupil dilation, and an increased heart rate (tachycardia) followed by a very slow heart rate (brachycardia). Coniine also *blocks* receptors in the central nervous system (the spinal cord), resulting in a creeping paralysis that starts with the feet and legs and spreads throughout the body. Initially there will be numbness in the legs and difficulty walking. Death, which may occur several hours later, is as a result of paralysis of the muscles responsible for breathing.

The specific toxicity of all of the *Conium* alkaloids has not been studied, but it seems that the shape of the molecule is particularly important. A propyl group (a chain of three carbon atoms) attached to a nitrogen-containing ring structure is needed to confer toxicity. Coniine is also a chiral compound (see page 55), meaning it can exist in two forms (or hands) with identical chemical composition but that are mirror images of each other. In the case of coniine, one form (the left-handed or *l*-form) is nearly twice as toxic as the other and is the form that dominates in hemlock plants. The synthetic coniine produced in 1886 would have been an equal mixture of both forms of coniine.

The consequences of not fully investigating biological interactions of the pure forms of a chiral compound were only realised with the thalidomide disaster of the 1950s. Thalidomide, a chiral compound, was prescribed as a mixture of both left- and right-handed forms, to treat morning sickness. Though it was effective in treating this, one hand of thalidomide (the left one) is a teratogen; it caused serious malformations in the babies that were born to mothers taking the drug, and thalidomide was withdrawn from the market in 1961.

The paralysing and numbing effects of coniine suggest that it could be used in medicine, and in the past it has been used to treat asthma. Coniine salts are described in the 1912 book *The Art of Dispensing* as being 'rarely prescribed' for the treatment of asthma. This was one of the books Agatha Christie would have studied for her dispensing exams in 1917. Coniine was also listed in the *British Pharmacopeia* as a sedative and antispasmodic, which led to its recommendation as an antidote to strychnine poisoning; it would have been administered in capsules or tonics. Pure coniine is an oil, which is reasonably soluble in water, but solids are preferable to work with when weighing and mixing drugs for prescriptions, especially when the oils are volatile and the vapour produces smells like mouse urine. Oils such as coniine might be converted to a salt for ease of use and storage in medical prescriptions. Coniine was usually prescribed as its hydrobromide salt, in doses of 1/100 or 1/60 of a grain (approximately 1mg). The drug's use was already waning by 1912, and coniine had completely disappeared from the *Pharmacopeia* by 1934; the therapeutic dose was found to be too close to the toxic one to be practical.

The way coniine relaxes muscles may still be of interest in medical research, so the physiological interactions of both its chiral forms are important to understand. Coniine would not currently be suitable for use in medicine without some modifications, as the compound does not bind selectively to receptors in the central nervous system, resulting in side effects. Modifications to coniine may be able to remove its teratogenic effects, too. This modified drug, whatever it might be, has potential applications in surgery.

Is there an antidote?

Treatment for hemlock poisoning is the same today as it would have been in the 1920s, when Amyas Crale is poisoned in Christie's *Five Little Pigs*, owing to the fact that there is no antidote. A stomach pump and treatment with activated charcoal are required to prevent more of the poison being absorbed from the stomach into the bloodstream; artificial respiration can then be used to support the patient while the body's metabolic processes eliminate the poison. This may take two or three days, but a patient receiving this treatment can expect to make a full recovery.

Agatha and hemlock

Coniine and its toxic properties are discussed in some detail in Agatha Christie's *Five Little Pigs*. Christie used the character of Meredith Blake, an amateur chemist who dabbles in herbs, to give the five suspects all the information and opportunity necessary to poison Amyas Crale. Five suspects were at the home of Amyas and his wife Caroline at the time of the murder. They were Philip Blake, a friend of Amyas; Meredith Blake, Philip's brother who lived nearby; Elsa Greer, a spoilt society beauty whose portrait Amyas was painting, and with whom he was having an affair; Angela Warren, Caroline's younger sister; and Cecilia Williams, Angela's governess.

Amyas and the rest of the Crale household visited Meredith the morning before Amyas's death. Meredith gave a tour of his laboratory, and pointed out various preparations he had made from plants. He took particular time discussing his preparation of coniine, which, he told the assembled audience, he had extracted from spotted hemlock. Meredith went on to describe the properties of coniine, and lamented the fact that it had disappeared from the *Pharmacopoeia* – Agatha Christie clearly kept up to date with changes in pharmaceutical practice. Meredith had found coniine to be effective in the treatment of asthma and whooping cough; coniine may relieve the symptoms of these, by numbing the pain and relaxing the muscles, but it would not treat the underlying conditions.

Meredith also read out a passage from *Phaedo,* where Plato describes the death of Socrates.

Agatha Christie goes into some detail when she describes the death of Amyas Crale. He was found slumped in front of his easel, sprawled on his seat with his arms flung out. He appeared to be staring at the picture he was painting. It was a position his friends and family had often seen him in, and those who had observed him from a distance that day didn't think anything was amiss. When it was finally realised that something was seriously wrong, no one could be sure that he was dead, so a doctor was called. 'He looked – almost natural. As though he might be asleep. But his eyes were open and he'd just stiffened up.' The doctor duly arrived, but it was too late.

Amyas's death might have been taken as natural causes, perhaps due to sunstroke, if it had not been for Meredith. On the morning of the murder, Meredith had noticed that the bottle containing his preparation of coniine was almost empty, when the day before it had been almost full. Concerned that someone might have taken the coniine without realising how dangerous it was, Meredith went to Amyas's house, to talk the matter over with his brother Philip. As he walked up towards the house he waved to Amyas, who had set up his easel in the garden to paint a portrait of Elsa Greer. Meredith watched Crale walk back to his easel and saw him stagger a bit. He thought Amyas was drunk, but in fact the poison had already begun to take effect.

Coniine leaves no distinctive signs at post-mortem, but the disappearance of the poison from Meredith's laboratory would have given the pathologist a clue as to what to look for. Enough poison was extracted from Crale's remains, probably using the Stas method, to prove that he had died of coniine poisoning. Once extracted from the body coniine could be easily identified, not only by smell but by characteristic chemical colour tests. These colour tests were crude, and unreliable by today's standards, but they would have left little doubt for pathologists and jury members in the 1920s. Today, coniine could easily be identified using chromatographic techniques. Even if coniine is not specifically tested for as part of a post-mortem examination, it would be picked up by more general tests for plant alkaloids, which are standard procedure in forensic toxicological screenings.

The doctor in the Crale case believed the poison had been administered two or three hours before the body was discovered. From the post-mortem findings it was easy to reconstruct the final hours of Amyas Crale's life. Agatha Christie was well aware of the symptoms of coniine poisoning, and describes the paralysing effects taking over Amyas's body. After the others had gone to lunch he would have dropped down and relaxed in his seat. Muscular paralysis would then have set in. Agatha trusted Plato's description of the death of Socrates saying that there was no pain, but as we've seen this may not have been the case. Amyas may have been in *considerable* pain, and would certainly have been distressed at his growing physical incapacity; the poison would have kept him conscious and alert until his final moments, though unable to call out for help.

In front of Amyas's body an empty glass and beer bottle were found, and they were taken away for analysis. Initially it was thought Amyas might have committed suicide. He had argued with his wife, but this was not an unusual occurrence and, therefore, there appeared to be no motive for him to take his own life. As the circumstances of the day were investigated by the police, suspicion fell on Caroline, Amyas's wife. An empty bottle of jasmine scent was found in a drawer in her bedroom. Analysis of the bottle revealed traces of coniine hydrobromide. Meredith must have prepared the hydrobromide salt of coniine in his laboratory, and stored it in a solution. It was thought that Caroline decanted some of the coniine into her emptied scent bottle when she visited Meredith's laboratory, while no one was looking. She claimed she had taken the poison to commit suicide, but she was not believed. In addition, it was Caroline who took the bottle of beer to her husband. It was assumed she had added the coniine to the beer with a pipette normally used for refilling fountain pens; the pipette was found crushed to splinters on the path between Amyas's easel and the house. This instrument would hold approximately 1–2ml of liquid. Assuming Meredith had produced a fairly concentrated solution of coniine hydrobromide, the pipette would have had more than enough capacity to carry a lethal dose.

Caroline was arrested and tried. She put up no defence, was found guilty and was then hanged. Many years later, it is up to Poirot to find out if Caroline really was guilty of the murder. He interviews the 'five little pigs', and reconstructs in his mind what happened on the fatal day. Of course, Poirot realises the true significance of many clues that the police investigation had overlooked, and is able to use his little grey cells to discover the truth.

There is no question that Amyas died of coniine poisoning, and it seemed reasonable that the coniine in his body was the same as that taken from Meredith's laboratory. But precisely who took it, and who administered it to Amyas, was less clear. The case against Caroline was stark; no one seemed in any doubt that she was responsible. One fact, however, proves to Poirot that Caroline could not have been the murderer; analysis of the beer bottle and glass found in front of Amyas revealed coniine only in the glass. Caroline had taken her husband the beer but, so far as anyone knew, she had not touched the glass. Another fact in Caroline's favour was that the poison must have been given to her husband before she brought him the beer. The time taken for the poison to act and the evidence from Amyas himself indicated that he had already tasted the bitter flavour of coniine before his wife arrived – when he drank the beer he commented that 'everything tastes foul today'.

But if Caroline wasn't guilty, who was? Anyone could have taken the poison from the laboratory, or observed Caroline taking it, then stolen it in turn from the scent bottle she had hidden in her drawer. Caroline believed her younger sister Angela had taken the poison, and put it into a beer bottle for Amyas to drink. Angela had a row with Amyas on the day of the murder, and was seen fiddling with a bottle of beer that

Caroline took from her and gave to Amyas. When his body was found Caroline was observed wiping fingerprints from the beer bottle and wrapping her husband's hand around it, in an attempt to suggest suicide and divert suspicion from Angela; she had not known that the coniine was only in the glass. Believing Angela to be guilty and wishing to protect her, Caroline offered no defence when she stood trial for murder. Poirot is therefore able to eliminate both Caroline and Angela from his list of suspects. But that still leaves four little pigs ... and to find out 'whodunit' you will have to read the book.

4.50 from Paddington

… Learn this, Thomas,
And thou shalt prove a shelter to thy friends,
A hoop of gold to bind thy brothers in,
That the united vessel of their blood,
Mingled with venom of suggestion,
(As, force perforce, the age will pour it in)
Shall never leak, though to do work as strong
As aconitum or rash gunpowder.

William Shakespeare, *King Henry IV Part 2*

The title of Agatha Christie's 1957 novel *4.50 from Paddington* (entitled *What Mrs McGillicuddy Saw* in the United States) refers to a journey taken by Mrs McGillicuddy after a shopping trip to London. During the journey two trains travel alongside each other, and for a brief moment they move at the same speed. Mrs McGillicuddy looks through the window to the carriage opposite, and sees a woman being strangled. She believes she has witnessed a murder, and goes on to tell the station-master and the police. With no missing person being reported and no corpse discovered, no one believes the old lady. Only Miss Marple is convinced that her friend has witnessed a crime, and she resolves to investigate further.

The first task is to find the body. A curve in the train track a little further along the route encloses a large estate, Rutherford Hall, the residence of eccentric biscuit baron Luther Crackenthorpe. The train embankment slopes down into the grounds, providing a convenient spot for the disposal of a body from a moving train. Miss Marple asks her friend, Lucy Eyelesbarrow, to act as housekeeper at Rutherford Hall, so she can root around in the shrubbery for a dead body when no one is looking. The discovery of a recent corpse hidden in a sarcophagus in an outbuilding is the start of a murderous campaign against the Crackenthorpe family; two more characters are bumped off before Miss Marple can solve the crime.

Christie used two notorious poisons to dispatch members of the Crackenthorpe family: arsenic and aconitine, poisons that have been used to murder for millennia. But while arsenic is still well known and is used almost as a by-word for poison, aconitine has been more or less forgotten. By the 1950s aconitine's time in the limelight was almost up, but it had not disappeared completely. Aconitine-containing plants grow wild in much of the northern hemisphere, and several species are cultivated in gardens. Some of these plants have been given saintly names such as monkshood, but they all have a dark and sinister side.

The genus Aconitum

Aconitum variegatum – monkshood – is considered to be the most poisonous plant in Europe; it has been called the 'Queen Mother of Poisons'. The genus *Aconitum* occurs across the northern hemisphere, often in mountainous regions. All these plants contain aconitine, an alkaloid, in addition to a number of related compounds. The genus contains around 250 species, of which some of the common names include monkshood (named after the shape of its flowers), wolfsbane, leopard's bane and Devil's helmet. 'Bane' means 'poison', and this refers to the use of the plant as an arrow-poison for hunting wolves and other dangerous carnivores. As these names suggest, these plants have something of a reputation. In Ancient Greece they were believed to have been spawned from the drool of the hound of hell, the ferocious three-headed dog, Cerberus. Hercules was challenged to bring the hell-hound from the Underworld to the surface. As he wrestled with the beast, saliva from its three mouths scattered over the rocks; where the drool fell, the poisonous flowers grew.

Preparations of *Aconitum* plants have been used in witchcraft medicine for centuries for the treatment of gout, probably as an analgesic to ease the intense pain of the condition. The roots of the plants were also a common ingredient in witches' flying salves; alkaloids within the roots have a 'local anaesthetic' action, giving a feeling of numbness, and this may have contributed to a sensation of the body losing contact with the ground.

The use of *Aconitum* species and their extracts persisted in medicine until the early twentieth century. They were formerly used in drops to reduce heart rate, fevers and elevated blood pressure, and to promote sweating. Extracts could also be applied externally as a liniment for the relief of pains such as neuralgia, rheumatism, sciatica, migraine and toothache. The numbing effect produced by the alkaloids would have relieved localised pain, but the margin between a dose producing numbness and one causing serious toxic effects is dangerously narrow. The safety margin was too narrow for a Dr Meyer who, in 1880, prescribed aconitine drops to a young boy. After treatment the boy became ill with chills and convulsions, and his mother went back to see the doctor, blaming the medicine for the child's illness. Dr Meyer was so outraged at someone daring to question his prescription that he took a dose from the boy's medicine bottle to prove that it was perfectly safe. Five hours later Dr Meyer died of aconitine poisoning.

The high toxicity of these compounds means there are no modern medical uses for either the plants or the individual alkaloids they contain. However, aconitine is still used in Chinese traditional medicine as a pain-relieving analgesic, and for its anti-inflammatory properties. The roots are processed by soaking and boiling to reduce their toxicity before use. Poisonings still occur, either because more than the prescribed dose has been consumed, or because the processing of the root has not been carried out effectively. Most cases of aconitine poisoning in recent years have taken place in Japan and China. These cases are usually accidental or suicide. Murder cases are exceptionally rare, but not unheard of. To find a case of large-scale poisoning we have to look back to 1857, during the Indian Mutiny. Regimental chefs added *Aconitum* root to a soup that was to be served to a detachment of British officers. When the chefs refused to taste the soup one of the officers, John Nicholson, force-fed some of the food to a monkey, which died immediately. The chefs were hanged without trial.

Alkaloids that can be extracted from *Aconitum* roots include mesaconitine, hypaconitine and jesaconitine, as well as aconitine. The concentration of alkaloids and the ratios of the different forms may vary enormously depending on the species of plant, the region and the season. Most of the poisonous compounds are found in the roots, but all parts of these plants can be lethal if ingested. This happens more often than you might expect, because the roots are often mistaken for horseradish. A tragic case from 2004 illustrates the confusion: Andre Noble, a young Canadian actor, went hiking with his aunt and ate *Aconitum* roots, mistaking them for radishes. He fell ill at his aunt's cabin, and died on the way to hospital.

How aconitine kills
As we've seen, aconitine is one of a number of alkaloids that occur in *Aconitum* plants; it was once used in medical applications, and was the poison used in *4.50 from Paddington*. Aconitine is poorly soluble in water but it readily dissolves in fats and oil; its solubility in fats increases its absorption across

the skin, which allowed the compound to be used in creams and salves for external application. It also means that toxic effects can be experienced by gardeners who handle the plant without wearing gloves.

When aconitine has been absorbed into the bloodstream it is distributed throughout the body, but it binds preferentially to sites that form part of sodium ion channels, found in the cell membranes of nerve and heart cells. Sodium channels allow sodium ions (Na^+) to flow into a cell, triggering the release of potassium ions (K^+) out of it. Switching the positions of the sodium and potassium ions is a process called depolarisation. Rapid depolarisation results in an electrical signal that is transmitted along the length of a nerve cell as a wave of sodium channels opens along its length. In heart cells the movement of sodium ions triggers the contraction of the cell, and the coordinated contraction of these cells results in a heartbeat (see page 97). After each electrical signal or contraction the cell must reset itself to allow the process to be repeated, and molecular pumps move the sodium and potassium ions back to their original positions.

Aconitine is an agonist; it binds to a site on the sodium ion channel and activates it. The channel opens and sodium ions flood in, causing the nerve to fire or the heart cell to contract. But the aconitine causes the channel to stay open and the cell to remain depolarised. The cells cannot reset to their original arrangement because of the increased in-flow of sodium ions – like trying to empty a bath with the taps still running.

The result of aconitine's interactions with nerve and heart cells can be seen almost immediately; symptoms are rarely delayed for more than an hour. Aconitine affects sensory and movement responses as well as the heart, creating an irregular and rapid heartbeat. It produces a wide range of effects and sensations, a few of which appear to be unique to aconitine. A burning sensation is often experienced, which feels like a red-hot poker being drawn across the tongue. There is a sensation of numbness and tingling in the mouth and throat, and a feeling that the throat is swelling up. The individual may also experience giddiness and a loss of muscular power. The pupils become dilated, the skin grows cold and the pulse feeble, with laboured breathing and 'a dread of oncoming death'.* Finally numbness and paralysis set in, rapidly followed by death in a few sudden gasps. Death, due to paralysis of the heart or breathing, usually occurs within two to six hours, but with very large doses death is almost instantaneous. A dose as low as 1–2mg can be fatal. This is a serious poison.

Non-lethal doses of aconitine can be excreted by the body, but the time taken can vary hugely between individuals. The half-life of aconitine is generally in the range of four to 24 hours. Excretion is mainly via the kidneys (though it can also occur via faeces). The rate at which it leaves the body depends on how efficiently the kidneys can filter the compound from the bloodstream and into the bladder. This filtration and excretion process can be affected by the efficiency of the heart; if aconitine is impairing its normal function, the rate of elimination will be slower. Aconitine may also affect the health of the kidneys, further slowing the rate of elimination from the body.

Is there an antidote?
There is no specific antidote available to treat aconitine poisoning, even today. The best care that can usually be offered is prevention of further absorption of the poison. This can be achieved by giving activated charcoal to absorb the poison, and using a pump to remove unabsorbed toxins from the stomach. Supportive care can then be given, such as artificial respiration to maintain breathing, and drug treatments to normalise the heart rhythm. If the patient survives the first 24 hours they can be expected to make a full recovery.

Some experimental antidotes for aconitine poisoning have been discovered, occasionally by accident, but none is a recognised or recommended method of treatment. One

*This pleasing quote is from *Poisons and Poisoners* by C. J. S. Thompson.

involves giving the patient lidocaine, a local anaesthetic often used by dentists. The drug works by blocking the sodium ion channels in the nerves, thereby counteracting the effects of aconitine. This treatment was given to a Japanese man admitted to hospital with aconitine poisoning in 1992. He was suffering from premature ventricular contraction and ventricular tachycardia – heart problems that can prove fatal if not treated quickly. After administration of lidocaine the patient's heartbeat returned to normal. This drug was first synthesised in 1943 by Swedish chemist Nils Löfgren (1913–1967), and his colleague Bengt Lundqvist (1922–1953) performed the first anaesthesia experiments by injecting himself with it.

A more unusual treatment was discovered in 1992. A paper in the *Journal of Experimental Medicine* discussed the case of a 33-year-old woman who vomited and collapsed in the lobby of a Japanese hotel. An ambulance was called, but the woman fell unconscious during the journey to the hospital and died shortly after arrival. Her death was due to ventricular fibrillation, what is commonly thought of as a heart attack. A post-mortem analysis of her blood revealed the presence of aconitine, mesaconitine and hypaconitine, and it was determined that she had died of poisoning (through ingestion of an *Aconitum* plant, rather than the pure aconitine compound). Further analysis revealed that tetrodotoxin was also present in the woman's blood; this is a poison found in the skin, ovaries and liver of pufferfish. The raw fish, carefully prepared so as not to contaminate the flesh, is known as *fugu*, and it is a great gastronomic delicacy in Japan. Tetrodotoxin acts by inactivating sodium ion channels in nerve cells, and causes death by paralysis of the diaphragm. Tetrodotoxin binds to a different site in the channels to aconitine, and it can delay its action when the two substances are administered in a particular ratio. The symptoms and death of the woman in the Japanese hotel were delayed by the presence of this pufferfish poison.

Pufferfish poisoning has been known of for centuries. Captain Cook recorded the first known case in 1774, when some of his crew ate pufferfish and the leftovers were given to the pigs. The crew suffered from shortness of breath; the pigs died. The active ingredient of pufferfish poison was first isolated in 1909 by Yoshizumi Tahara, and it has been used since the 1930s in Japan to manage pain in terminal cancer and for migraines.

Some real-life cases

Murders involving the use of aconitine or *Aconitum* plants are very rare, though there has been a relatively recent case. In 2009, Lakhvir Kaur Singh attempted to kill her ex-lover, Lakhvinder 'Lucky' Cheema, and his fiancée, Gurjeet Choong, using curry laced with poison. The announcement of the couple's engagement was too much for Singh, who apparently bought her poison – probably a crude extract from *Aconitum ferox*, a species known as Indian aconite, and one of the most poisonous plants in the world – especially for the purpose of poisoning the two. After eating the curry, Lucky and Gurjeet complained of feelings of numbness, pains in the stomach, failing sight and dizziness. An ambulance was called but doctors at the hospital were unable to save Lucky's life. Gurjeet recovered after being placed in a coma for two days while the poison was identified and treatment administered. The murderer received a sentence of a minimum of 23 years' imprisonment.

The best-known case of aconitine poisoning, one that Agatha Christie was almost certainly aware of, occurred in 1881. There are several similarities between the real-life case and the fictional poisoning written about later by Christie. Dr George Henry Lamson was a medical doctor who volunteered as an army surgeon in Romania and Serbia. When he returned to England he set up a medical practice in Bournemouth, but he had acquired a morphine habit, perhaps as a result of his experiences of war. Initially Lamson prospered, but as his morphine addiction took over his life his practice floundered, and debts started to build up.

Financial relief came in 1879 in the form of an inheritance. Lamson's wife, Kate, was one of four siblings who had equal shares of the inheritance from their parents. When Lamson's brother-in-law Herbert John died, his portion of the inheritance was redistributed amongst the three remaining siblings.* However, Lamson's financial relief was short-lived, and his debts continued to grow.

Lamson decided that the only way out of his financial difficulties was another inheritance, and he set his sights on Percy John, Kate's 18-year-old crippled brother. Percy had a curvature of the spine that had paralysed him from the waist down, though he had full use of his upper limbs and was in otherwise good health. Lamson made his first attempt on the boy's life in the summer of 1881. While on holiday on the Isle of Wight, Lamson gave Percy a pill that he dutifully swallowed. Soon afterwards he became very ill, but he made a complete recovery and returned to his boarding school in Wimbledon for the autumn term. Lamson's money worries were becoming acute, and he went to America to try to make his fortune but returned with his situation unimproved. While he was there, Lamson made a significant purchase; a type of gelatin capsule designed for taking powdered medication.

On 24 November 1881, Lamson made another important shopping trip. He bought two grains of aconitine (approximately 130mg) from a pharmacist in London. Lamson was unknown to the pharmacist, but because he was a medical doctor he was able to purchase poisons without having to answer any awkward questions – the pharmacist simply asked for Dr Lamson's name and checked in the register of medical professionals. Finding everything in order he sold the aconitine to Lamson for 2s 9d.†

On 3 December, Lamson paid a visit to Percy at his school in Wimbledon. When he arrived he sat down to talk with

---

*A similar division of the spoils of inheritance crops up in *4.50 from Paddington*.

†14 p in new UK money (around 20 US cents); this is the equivalent of around £12.50 ($19) today.

Percy and the headmaster. Sherry was served, to which Lamson added a spoonful of sugar, claiming that it counteracted the effect of the alcohol. At some point during the visit Lamson produced a Dundee cake with three slices already cut and he proceeded to offer them to Percy and the headmaster. He took the last slice for himself. Conversation turned to Lamson's recent trip to America, and he produced some of the capsules he had bought there. He recommended the capsules to the headmaster as a means of giving bitter medications to the pupils so they wouldn't have to taste them. To demonstrate, he filled one of the capsules with sugar, from the same bowl he had used for his sherry, then pushed the two halves together. He gave the capsule to Percy, complimented him on being a champion pill-taker, and asked him to show the headmaster how easy it was to swallow these special pills. Percy did as he was told. Lamson then promptly made his excuses and left, saying he did not want to miss his train to catch the boat to France. In fact he had just missed one train and the next would not be for 30 minutes or so; the station was just a few minutes' walk away, but Lamson would not stay.

Within ten minutes of Lamson leaving, Percy became ill. He vomited and complained of stomach pains. He was carried up the stairs to his room by his friends. He said he felt the way he had after taking the pill Lamson had given him on holiday. His condition worsened, with his whole body convulsing so he had to be forcibly held down. Two doctors attended the boy. Both were baffled by Percy's symptoms, though there was no doubt he was in considerable pain. His mouth and throat were burning; he said it felt as if his skin was being pulled from him. Not surprisingly, Percy was writhing in agony. The doctors administered two doses of morphine in an effort to relieve the pain, but they were at a loss as to how to help the boy. The doctors later admitted in court that they knew nothing about the action of a fatal dose of aconitine on the human body. Percy died that night, after suffering four hours of torment.

The doctors believed that the boy had been poisoned with some kind of vegetable alkaloid. Suspicion fell on Lamson almost immediately, and the police began to search for him. Though he had successfully made it to France, Lamson voluntarily returned to England, and walked into a police station to help with their enquiries. He was promptly arrested for murder.

A post-mortem examination of Percy's body had been ordered, but no obvious signs as to the cause of death could be detected. Dr Thomas Stevenson (1838–1908), an expert in alkaloid poisons, was brought in to examine the remains. He managed to extract a substance from Percy's organs, but there was no chemical test to identify aconitine (and there still isn't). Stevenson had to rely on his extensive knowledge of the taste of alkaloids. The doctor had a collection of 50 to 80 different alkaloids in his laboratory, and he could identify all of them by taste; his party trick was to identify a particular alkaloid by taste before his colleagues could complete the chemical test to confirm its identification. The taste and burning sensations of aconitine were, he claimed, unique. He proposed that as little as 1/60 grain of aconitine, which equates to approximately 1mg, could prove fatal.

Lamson was probably well aware that there was no known chemical test for aconitine, and he had chosen this poison deliberately. Lamson had learned about aconitine when he was a medical student studying under Robert Christison (see page 113), Professor of Medical Jurisprudence at Edinburgh University and respected toxicologist, who had given evidence in several poisoning cases in Scotland. Lamson's defence did their best to throw doubt on the scientific evidence, as so little was apparently known about aconitine poisoning. The pharmacist who sold the aconitine to Lamson had also come forward to testify. Despite keeping no record of the transaction (he was not required to by law) the sale was so unusual that it stuck in his mind, and when he later read about the poisoning case in a newspaper he contacted the police. Another damning piece of evidence was found in Lamson's notebook, where he had jotted down the symptoms of aconitine poisoning.

To this day no one knows exactly how Lamson administered the poison, though it seems likely it was in either the pill or the cake. A lethal amount of aconitine could have been present in the pill capsule while still leaving plenty of room for it to be filled with sugar. An alternative theory, worthy of Dame Agatha herself, is that the poison was in a raisin in the slice of Dundee cake given to Percy. Despite not knowing precisely how he had carried out the crime, the jury took just 30 minutes to find Lamson guilty, and he was sentenced to death.

Lamson's time in prison forcibly broke his morphine habit; perhaps his newly acquired lucidity made him realise the cruelty of his actions. Four days before he was executed he confessed to the murder of Percy John.

Agatha and aconitine

In *4.50 from Paddington* Agatha Christie uses aconitine to bring an end to Harold Crackenthorpe's life. With Harold out of the way his siblings stand to inherit a greater slice of the Crackenthorpe family estate after their elderly father dies. The method of murder is straightforward. Tablets are sent to Harold Crackenthorpe, ostensibly from his doctor, Dr Quimper. Somebody takes a box normally used for sedative tablets prescribed to Emma (Harold's sister), substitutes aconitine tablets, and sends them from Emma's home, Rutherford Hall, to Harold at his London address. The chemist who made up the prescription could have made a mistake, but he claims no knowledge of the prescription at all. So someone must have got hold of some ordinary tablets and adulterated them with aconitine.

Getting hold of the few milligrams of pure aconitine necessary to kill an adult man would not have been easy, even in 1957. One method would be to extract the poison from an *Aconitum* plant. A relatively pure sample would be needed, to take up as little room in the tablets sent to Harold as possible, and to avoid altering their appearance. Although there is no mention of it in the book it would not be unusual to find monkshood growing in the grounds of Rutherford Hall, or in nearby countryside. Monkshood is a herbaceous perennial, dying back each winter to regrow the following spring from the roots. The poisoning occurred in the winter, so the murderer would have to know the location of the roots very well, or would have had to plan ahead and collect monkshood roots during the previous summer. Extracting the poison in a crude form would be relatively easy and could be achieved with some technical knowledge and little more than standard kitchen utensils. However, isolating aconitine from all the other alkaloids present in the plant would be more difficult, and would require more specialised chemical equipment. This seems convoluted and unlikely; it would be much easier to buy the already purified aconitine from a pharmacist.

The compound would have been stocked at a local pharmacy where prescriptions were made up to order. However, in 1957 aconitine was rarely prescribed and, as Miss Marple explains, 'they were the kind of tablets that are usually kept in a poison bottle, diluted one in a hundred for outside application'. The elderly spinster consistently displays a worryingly detailed knowledge of pharmaceuticals and poison.

Anyone wanting to buy aconitine from a pharmacist would have needed a prescription from a doctor or, if the pharmacist was satisfied that the aconitine was for a legitimate use, the poison would be sold and the poison register signed by the purchaser. Obtaining aconitine was a problem addressed by Oscar Wilde in his short story *Lord Arthur Savile's Crime*. Wilde was writing in the 1880s, a few years before the Lamson case, and even then there were restrictions on the sale of this poison. In the short story Arthur Savile, a carefree young man recently engaged to Sybil Merton, has his palm read and is told he will commit a murder. He decides to kill someone before his wedding so that he can begin his married life without the burden of not knowing when or whom he might kill. After careful consideration he elects to kill an elderly aunt using poison. Lord Arthur conducts his research by reading a pharmacopoeia and *Erskine's Toxicology*, and finds 'a very interesting and complete account of the properties of aconitine, written in fairly clear English … It was swift – indeed, almost immediate, in its effect – perfectly painless.' *Erskine's Toxicology* does not exist, but Wilde would have done well to read a book on toxicology since, as we have seen, aconitine poisoning, though swift, is far from pain-free.

Lord Arthur calls at a pharmacy in London to purchase the pill he requires. His request is initially refused because a medical certificate is required, but Lord Arthur explains that the aconitine is needed to put down a large dog that has shown signs of rabies. This apparently satisfies the chemist and the prescription is handed over. Mercifully for the aunt, she dies before taking the aconitine, and Lord Arthur has to find another victim.

The poisoner in *4.50 from Paddington* must have convinced the pharmacist that the aconitine was a legitimate prescription. After the murderer obtained the pure aconitine compound they would have added a few milligrams to Harold's tablets. This would be easy if the tablets had been in the form of gelatin capsules (see page 149), a casing of two halves used to hold powdered forms of drugs that have been around since the 1840s. Because Harold swallowed the tablets straight down with a glass of water, that he would not have initially noticed any unusually bitter taste or burning sensation in the mouth.

The prescription says to take two tablets nightly. Harold had been taking the medication for a while, but thought Dr Quimper had told him that he no longer needed it. Assuming he has misunderstood the doctor Harold dutifully takes two tablets with a glass of water before going to sleep. His death is announced to the family the next morning.

No mention is made of any symptoms displayed by the victim. Perhaps Harold didn't display any. This could only happen if Harold fell asleep quickly after taking the tablets and received a large dose of the poison, killing him rapidly with little time for symptoms to display themselves. Perhaps Christie was saving her readers from a graphic description of the pain and suffering Harold may well have experienced.

The morning after Harold's death it has already been established that aconitine poisoning was responsible, though we are not told how this was determined. There would be no obvious signs on the body; even a post-mortem examination wouldn't have revealed any damage to organs characteristic of aconitine. Extracting aconitine, or indeed any substance, from body tissue takes time, and it would be easier to extract aconitine from the tablets Harold had been sent, even if it was mixed with other compounds in the tablets. But there is still the problem of identifying the substance that had been extracted. There has never been a colour chemical test available to identify aconitine. A brave or foolhardy pathologist might be able to quickly identify the poison by tasting it and recognising the characteristic burning sensation aconitine produces. Happily, by 1957 more reliable and less uncomfortable methods were being used to identify poisons. Chromatography methods would have been available at this time, although these rely on having a standard available to compare the results. Aconitine is such a rarely used compound, for medicine and murder alike, that it seems unlikely that an aconitine standard would have been readily available. A standard could have been produced, but this would have taken time. Animal tests were another option that was available, but this also would have been time-consuming.

While the victim is alive, blood and urine samples can be used to identify the poison. Even after the poison has disappeared from the blood it may still be traced in the urine because the toxin accumulates in the bladder until it is expelled. Urine is rarely available at post-mortem, so a pathologist would be relying on samples of the liver and kidneys, where the poison is more concentrated. Blood levels of aconitine tend to be very low compared to the levels in these organs. Even so, isolating a tiny amount of poison distributed throughout organs weighing many grams is not an easy task. Though the cause of Harold Crackenthorpe's demise could have been established it would probably have taken longer than suggested in the novel.

Aconitine is an unusual but effective choice of poison for Agatha Christie. There is very little discussion of the properties of aconitine in the book, and what *is* described is limited to its pharmaceutical preparations – the aspects with which Christie would have been most familiar. However, given the very limited number of murder cases she had to draw on, a few small mistakes – such as the time taken to detect the poison being far too short – are understandable.

Three Act Tragedy

Mr Babbington was peering across the room with amiable short-sighted eyes. He took a sip of his cocktail and choked a little. He was unused to cocktails, thought Mr Satterthwaite amusedly …
Mr Babbington took another determined mouthful …
'Look,' said Egg's voice. 'Mr Babbington is ill.'

Agatha Christie, *Three Act Tragedy*

Agatha Christie's 1935 novel *Three Act Tragedy* (entitled *Murder in Three Acts* in the United States) is the only one of her novels to use nicotine as the means of murder. The three victims – a mild-mannered vicar, an eminent doctor and a patient at a sanatorium – appear to have nothing in common. The first death is, after some consideration, initially attributed to natural causes; the second occurs under similar circumstances, with the victim displaying almost identical symptoms before dying – after which it is realised that there's a murderer about.

The third is carried out to silence a witness. All the victims are dispatched by a lethal natural product, nicotine. Among the suspects are an actor, a dressmaker, a playwright and even a butler, none of whom seem to have a motive. Fortunately, Hercule Poirot is on hand to sort through the red herrings and reveal the culprit.

Most people are aware that nicotine is dangerous. It is indirectly responsible for thousands of deaths annually through smoking; nicotine causes the addiction, but it is other compounds from tobacco smoke that usually kill smokers. Pure nicotine, however, is highly toxic in its own right, and has been the cause of many fatalities; it has rarely been used in murder, though. This is surprising given its ready availability, but perhaps because it is so commonplace we struggle to believe that such an everyday substance could be used to kill.

The nicotine story

At room temperature, nicotine is a clear colourless liquid. It mixes completely with both water and alcohol and has a strong, distinctive taste. When exposed to air it turns brown, almost whisky-coloured. The pure liquid gives off a characteristic tobacco-like smell.

The chemical is an alkaloid found in plants of the genus *Nicotiana*. These are part of the family Solanaceae (which we met on page 51) that includes plants such as deadly nightshade, the tomato and the potato. Nicotine actually occurs in all Solanaceae, but it is *Nicotiana* that has the highest concentrations of the compound. There are many species in the genus, which is native to the Americas, Australia, south-west Africa and the South Pacific, though they are commonly grown in gardens across Europe for ornamental purposes. *Nicotiana tabacum* is the species grown commercially, with tobacco products derived from the dried leaves of the plant.[*]

[*]All parts of the plants contain nicotine, but the highest concentration is in the leaves.

Nicotine can be absorbed through the skin, the lungs and the gastrointestinal tract, hence the wide variety of tobacco products that can be smoked, chewed, inhaled or attached to the skin (it is nicotine itself that is incorporated into patches, gum and e-cigarettes as the key ingredient used to help break the addiction to smoking).

Tobacco has been smoked, chewed and snorted for hundreds if not thousands of years in the Americas, and was first brought to Europe in 1528 by returning Spaniards. It was immediately popular; by as early as 1533, there is mention of a tobacco merchant trading in Lisbon. Both nicotine and *Nicotiana* derive their names from Jean Nicot (1530–1600), a French ambassador to Spain. In 1559 Nicot introduced tobacco to France, when he sent some seeds and dried leaves to the French royal court, along with information about the medicinal properties they were thought to possess. Tobacco became instantly popular with the royal family, and particularly with Catherine de' Medici, the Queen Mother. Fashionable members of Parisian society – always keen to imitate royalty – started using tobacco, and Nicot became a celebrity.

Dried tobacco contains between 0.6 and 3 per cent nicotine. Very little of this (around 1–2mg per cigarette) makes it into the bloodstream when cigarettes are smoked, because most of the nicotine is burnt. Consequently, the average smoker is highly unlikely to be exposed to toxic levels of nicotine. Smokeless tobacco products such as snuff or chewing tobacco have been shown to generate higher nicotine levels in the blood. The rate of absorption is slower with smokeless tobacco but more nicotine is available (because it is not being burnt), and therefore more nicotine will be absorbed. The nicotine released by chewing tobacco increases salivation, which is part of the reason why it was once popular with baseball players, as it kept the mouth moist while they were out on the dusty baseball field. Unwanted juices are spat out, but accidental swallowing has been known to give a fatal dose of nicotine.

Those working with tobacco plants or involved in harvesting or processing the leaves of the plant are at significant risk from absorbing nicotine through their skin; this gives rise to a condition called green tobacco sickness. This affects up to 89 per cent of tobacco harvesters over the course of a season. Nicotine is highly soluble in water so workers picking tobacco leaves when it is raining, or when the plants are covered in morning dew, can absorb nicotine through their hands. Clothes may not offer any protection if they are not waterproof. Sufferers may feel nauseous, dizzy or develop a headache. There may also be increased sweating, salivation and difficulty breathing. The effects usually wear off after a couple of days, but in severe cases there can be fluctuations in blood pressure and heart rate that may require emergency medical treatment. The symptoms seem to be less severe in workers who smoke, as their bodies have adapted to the presence of low levels of nicotine.

Children are more susceptible to nicotine poisoning than adults, owing to their lower body mass. There have been several cases of nicotine poisoning in children who have eaten a cigarette end; there have also been accidents with nicotine gum, nicotine patches and e-cigarettes. There has even been a case of severe nicotine poisoning from a traditional remedy for treating eczema. A Bengali remedy used tobacco leaves, coffee and lime powder, as well as prayers, to cure skin conditions such as eczema, ringworm and scabies. However, the broken skin on the child's arms, caused by the eczema, probably allowed the rate of absorption of nicotine to rise, and after just 30 minutes the child started to feel unwell (fortunately a full recovery was made).

The poisonous properties of nicotine have been known for almost as long as Europeans have been smoking it. Toxic effects are not restricted to humans: nicotine is poisonous to anything with a nervous system, hence its use as an insecticide since at least the sixteenth century. Huge quantities of nicotine were used as insecticides in the 1940s, because it was readily available as a by-product of the tobacco industry. Nicotine can be considered a 'green' insecticide for use on organic crops, because it is a natural substance. However, it is clear from accidental poisonings that this was a very dangerous product. Nicotine *can* look very much like whisky, and some gardeners have unwisely stored their insecticides in old whisky or cognac bottles. Fatal mistakes have been made when gardeners have reached for a bottle of booze and taken a swig of the wrong liquid. Since the 1940s the amount of nicotine in insecticides has been reduced, and it's been replaced by nicotine derivatives, or other compounds less harmful to mammals. Nicotine insecticides are banned in Europe and the United States nowadays.*

How nicotine kills

Nicotine is a very fast-acting poison; it can kill in as little as four minutes. Inhaling nicotine is the quickest way to absorb it into the body. The lungs have thin membranes between the alveoli (tiny air sacs) and bloodstream that enable oxygen and carbon dioxide to be absorbed or expelled easily. This also makes it easy for nicotine to cross into the bloodstream. Nicotine from cigarettes can reach the brain in seven seconds. Absorption through the skin takes around an hour, but this varies depending on the individuals' skin as well as the form of delivery (for example, whether the nicotine is from a patch or in the dew of a tobacco plant). Absorption of nicotine that's swallowed (rather than inhaled) mostly occurs in the mouth or intestines, because the acidic environment of the stomach prevents the absorption into the bloodstream.

Nicotine targets specific sites within the body, by acting on a subset of receptors found in nerves. As described earlier (see page 56), signals are transmitted across the gaps between two nerve cells (or between a nerve cell and a muscle cell) by chemicals called neurotransmitters. In 1900 the Cambridge

---

*Though their legacy lives on in the form of neonicotinoids, a group of chemicals very similar to nicotine that kill insects but have fewer side effects in vertebrates. Although they were originally designed to act on pests, these compounds also have a devastating effect on insects that are beneficial to plants, bees in particular.

physiologist John Langley (1852–1925) discovered that receptors in the muscles that respond to the neurotransmitter acetylcholine are also stimulated by nicotine, so they were named nicotinic receptors. Nicotinic receptors are found at junctions within the central nervous system (CNS) and where nerves meet muscles; they are also found throughout the autonomic (or 'below consciousness') system, where they increase activity in the sympathetic nervous system, responsible for the 'fight-or-flight' response to a perceived danger. Activation of these nerves widens the pupils of the eyes, increases the heart rate and dilates blood vessels to the heart, brain and muscles.

Stimulation of the nicotinic receptors in muscles can cause the muscle to contract, which is the cause of the twitching sometimes seen in new smokers. These receptors quickly become desensitised to the presence of nicotine, so the twitching stops with continued smoking. Heavy smokers may show a tremor in the hands, though, due to the higher levels of nicotine in their bloodstream. The body of a regular smoker will also adapt to the presence of nicotine by metabolising and excreting it more rapidly.

Nicotinic receptors are also found in the brain, and it is the stimulation of these nerves that is the root cause of nicotine addiction. Stimulation of the part of the brain known as the ventral tegmental area (VTA) gives rise to feelings of pleasure. In an experiment, rats had electrodes placed in their VTA, and were allowed to press a lever that triggered a current in them. The rats kept on pressing the lever to enjoy the pleasurable effects. Some rats kept on pressing the lever at the expense of eating and sleeping, and had to be physically removed from it to prevent them from dying of exhaustion. Chemicals produced in the brain, such as dopamine, naturally stimulate the VTA; therefore chemicals that make dopamine more available to the nerve cells of the VTA tend to be addictive. For example, cocaine stops cells retrieving dopamine after it has been released, so the dopamine is outside the cell for longer and available to interact with receptors. In the presence of nicotine, cells in the VTA release dopamine more readily than usual. Nicotine also reduces the amount of monoamine oxidases (MAO) in the brain to about half. MAO is a family of enzymes found throughout the body, one of which is responsible for the breakdown of dopamine. With less MAO present there is more dopamine around to interact with receptors in the VTA.

Receptors in the brain are more sensitive to nicotine than those in the rest of the body. However, a tolerance slowly builds up, and ever-increasing amounts of nicotine have to be obtained to yield the same levels of stimulation. Animals have been shown to become addicted to pure nicotine, so it is nicotine, rather than one of the other 3,000 chemical compounds found in cigarette smoke that is responsible for causing addiction. 'Addiction' is a difficult concept to quantify; scientists have looked at the physical effects of the drug in the body, as well as the consequences of withdrawal and the desire for another 'hit'. Although the physiological effects on the brain can be measured, there are many factors that contribute to addiction, both social and habitual. Cocaine and nicotine have been shown to have similar effects on the brain, but the experience of smoking is less pleasant than taking cocaine, suggesting that the addictive quality of nicotine is actually greater. However, some researchers comparing nicotine to cocaine concluded that nicotine was actually less addictive. What is not in any doubt is that cocaine and nicotine are both highly addictive substances.

Sufferers of Alzheimer's disease have fewer nicotinic receptors in their brain, which is thought to impair learning, reasoning and memory. Research has been undertaken into the use of nicotine patches to improve cognition in people with Alzheimer's. Smokers sometimes claim that smoking can increase attention, concentration and memory, and to a certain extent this is true. However, the raft of other harmful chemicals produced in cigarette smoke means smoking is a dangerous means of improving academic performance. Nicotine patches are perhaps a safer alternative but before you stock up on nicotine patches or bulk-buy nicotine gum for an exam, remember to check the dosage.* People have attempted suicide with nicotine patches, and several of them have required hospital treatment. An additional factor to consider is that the skin acts as a reservoir for nicotine, and it can continue to release it into the bloodstream hours after the patches have been removed. Serious poisoning incidents have occurred from the combined use of tobacco, nicotine gum and nicotine patches.

Another potential medical application of nicotine is for patients with schizophrenia. Schizophrenics are far likelier to smoke than other people, though the reason for this is not completely clear. It has been suggested that nicotine in the cigarettes helps control the symptoms of the disease, and schizophrenics who smoke are, in a way, self-medicating. Some schizophrenic-like symptoms can be induced in laboratory animals using drugs such as amphetamines. Activating nicotinic receptors in the brains of these animals has been shown to counteract some of the symptoms.

There can be serious side effects from non-fatal doses of pure nicotine. Harmful effects on the heart, blood pressure and muscles mean that nicotine is a somewhat risky treatment for both Alzheimer's patients and schizophrenics. Hopefully drugs will be developed that will interact preferentially with receptors in the brain, thereby reducing the side effects caused by nicotine acting on receptors in other parts of the body.

☼⚛

Nicotine has a dual effect, acting as both a stimulant and a depressant, depending on the dose. At low doses it is a stimulant; it stimulates the nicotinic receptors, causing nausea and

*Fans of the TV series *Sherlock* may remember Holmes using nicotine patches to help him with particularly difficult problems. This is not recommended.

vomiting, dizziness, headache, diarrhoea, an elevated heart rate (or tachycardia), an increase in blood pressure and sweating. Activation of nicotinic receptors in the brain causes an initial stimulation, alertness, a decrease in irritability or aggression, and a reduction of anxiety.

At high doses, nicotine becomes a depressant (with pain-relieving properties). There is an initial burning sensation in the mouth, throat and stomach, followed by a rapid progression to the symptoms seen in smaller doses. There may be convulsions, respiratory slowing, cardiac irregularities and coma. Death occurs up to four hours later (sometimes much more rapidly), and is due to paralysis of the respiratory muscles. If the patient survives for longer than four hours they will usually make a full recovery.

What exactly constitutes a lethal dose of nicotine for a human is a point of some debate. It is generally agreed that between 0.5 and 1mg/kg would constitute a lethal dose by injection or inhalation. This corresponds to a dose of between 40 and 70mg for a 70kg adult (one or two drops). A much higher dose would be needed to kill by absorption through the skin or by ingestion. A recent estimate of the lethal oral dose is between 500 and 1000mg (approximately ten to twenty drops).

Is there an antidote?
As is usual in an Agatha Christie novel, little was done to help the victims in *Three Act Tragedy,* though medical support might have saved them. The first thing to do is to remove the source of poisoning by washing the skin or enforced vomiting (emesis), depending on how the nicotine got into the body. It is likely that the patient will be vomiting anyway, and can remove much of the poison themselves. If the patient is not vomiting, activated charcoal can be given to absorb nicotine in the stomach, and a stomach-pump will remove more of it. Artificial respiration should be given as required.

There is, however, a specific antidote for nicotine – atropine (see page 49), which can be administered by injection. Atropine activates the nerves of the sympathetic nervous system, while nicotine, at high concentrations, will depress their activity. Additional measures can be taken to control the symptoms of nicotine poisoning as necessary; for example, anticonvulsants to control seizures.

Some real-life cases

Agatha Christie was quite right in *Three Act Tragedy* when she pointed out that nicotine was rarely used for the purposes of murder, and much of her knowledge of it must have been drawn from cases of accidental poisoning. But there is one famous case of murder by nicotine, dating from 1850. It is important, not only because it is an unusual choice of murder method, but also because it is the first case where scientific evidence was used to prove the presence of a plant-based poison in a corpse.

In a courtroom in France a few years before the murder in question, a prosecuting lawyer who was unsuccessfully trying to prove a case of murder by morphine declared thus: 'Henceforth, let us tell would-be poisoners ... use plant poisons. Fear nothing; your crime will go unpunished. There is no *corpus delecti* [physical evidence], for it cannot be found.' The fact that at the time nicotine was undetectable in a corpse may have been the reason why Count Hippolyte Visart de Bocarmé chose it as his poison, but it was probably his arrogance that made him believe he would never be convicted.

Count Bocarmé had an extraordinary life to match his extraordinary name. He was born during a thunderstorm in 1818 on board a ship bound for Java, where his father had been appointed Governor. The Count spent his early years in Java before returning to Europe with his family. He was a badly behaved young man, known to be a swindler and womaniser. When he was 24 his father died, and Hippolyte inherited his father's title and the family estate, Château de Bitremont near Bury, Belgium.

The inheritance was soon gone and, desperately short of money, the Count married Lydie Fougnies, the daughter of a retired grocer, believing her to be rich. Although she brought with her a small annual income, which more than doubled a few years later when her father died, it was nowhere near enough to support the excessive lifestyle the couple were leading. In order to fund the wild parties, extravagant hunts and their family of four children as well as a large household staff, they started selling off land. When this supply of cash ran out they started to look at Lydie's brother, Gustave Fougnies, in a new light. Gustave was unmarried, and had inherited the bulk of his father's fortune. He also suffered from poor health.* In his will Gustave considerately left everything to his sister upon his death. The Count and his wife assumed that they wouldn't have long to wait until they inherited the cash, so they continued with their expensive lifestyle, and mortgaged everything they could to fund it.

When Gustave announced he was getting married, the Count feared that his brother-in-law would change his will in favour of his new wife. He decided action was required before he lost the inheritance he felt he was due. By the beginning of 1850 Count Bocarmé had developed an intense interest in chemistry, and started a correspondence with a professor of chemistry, using a false name. With the knowledge he gained from the professor the Count was successfully able to distil a quantity of pure nicotine from a large amount of tobacco leaves that he had purchased during the summer of 1850.

On 20 November 1850, Gustave accepted an invitation to dinner at Château de Bitremont, during which he died. Only three people were present in the room at the time – Gustave, the Count and the Countess. The Count and Countess asserted the cause of death was 'apoplexy' (i.e. a haemorrhage) but the presence of bruising and scratches on Gustave's face indicated otherwise. Something had been forced into Gustave's mouth, and whatever it was had run down from the corner of his mouth, causing blistering to the skin.

---

*And, by the time of his death, he was recovering from a leg amputation.

If the marks on the body were not enough to arouse suspicion, then the behaviour of the Count and Countess immediately after the death certainly was. The Count tipped glass after glass of vinegar into Gustave's mouth. The body was also washed with vinegar, and Gustave's clothes were removed and taken to the laundry along with those of the Count and Countess from that evening. The Countess then busied herself with washing the floor in the dining room. Later, the Count applied himself to scraping the wooden floor of the dining room with a knife. The cleansing and tidying up continued until the afternoon of the following day, when the Count and Countess went to bed exhausted. Not surprisingly, the servants were very suspicious, and decided to call the authorities.

When a magistrate arrived the Count was reluctant to show him Gustave's body, and refused to pull the curtains back to allow him to see properly. He tried to shield Gustave's face with his hand but to no avail. It was apparent from the cuts and bruises that Gustave had not died a natural death.

Further investigations revealed inflammation in Gustave's throat and stomach, and it was concluded that he had been forced to drink some kind of corrosive substance, such as sulfuric acid, and that had been what killed him. Tissue samples from Gustave's body were bottled in alcohol and hastily taken to the laboratory of Jean Stas (1813–1891), with the request that he try to identify what had been used to kill Gustave. Stas was the most famous chemist in Belgium, and world-renowned for his work on atomic weights; he had converted his whole house into a working laboratory for his experiments.

A quick examination of the inflamed tissues in Gustave's mouth and throat convinced Stas that sulfuric acid had *not* been used. The damage from an acid would have been quite different. Like many other chemists at the time Stas made use of his senses of taste and smell in his experiments. He noted a taste of acetic acid in the remains, and the police explained how Bocarmé had doused the body in vinegar (the principal component of which is acetic acid), and poured many glasses of the stuff down Gustave's throat. Acetic acid alone would not kill a man, so Stas suspected the vinegar had been used to disguise the presence of another poison.

The eminent chemist worked night and day to extract whatever it might have been that killed Gustave. He added more alcohol to a portion of the remains, filtered it, added water and filtered it again. After evaporating off all the alcohol and water Stas was left with a sticky residue, to which he added caustic potash (potassium hydroxide, KOH). For the briefest moment, Stas smelled the distinctive aroma of nicotine.

Stas then spent three months developing a reliable method of extracting plant alkaloids from human tissue. The first step was to digest the tissues to release the alkaloid. This was done using acetic acid and alcohol. Gustave's murderer had already helped this process along by washing the body with vinegar, and the investigating authorities had helped Stas further by preserving the tissue samples in alcohol. The poison, now released from the tissues, would be dissolved in the alcohol. Stas reasoned that compounds within the body might be soluble in water or alcohol or neither, but not both. Nicotine (and other plant alkaloids), on the other hand, was soluble in both water and alcohol. By using a series of extractions with both these liquids, nicotine could be separated from the compounds normally found in the body. The final step was to wash the alcohol layer with portions of ether, and allow the ether to evaporate in a dish. What was left in the dish was a brownish residue with the unmistakable smell of nicotine.

Next, Stas carried out an extensive series of chemical tests to prove beyond doubt that the substance he had isolated was nicotine. He then contacted the police and suggested they look for evidence that Bocarmé had extracted nicotine from tobacco leaves. A thorough search was conducted at Château de Bitremont, and the chemical glassware that the Count had used was found hidden behind some wooden panelling, while in the garden they found the bodies of cats and other animals that Bocarmé had tested his tobacco extracts on. The gardener also remembered that the Count had purchased a large quantity of tobacco leaves the previous summer; he had told the gardener he was making perfume.

While the search of the château was going on Stas had continued his experiments, and he had extracted enough nicotine from Gustave's liver and tongue 'to kill several persons'. He also analysed clothing and wood shavings from the floor at Château de Bitremont to determine the presence of nicotine. In another experiment, Stas killed two dogs by administering nicotine by mouth. One dog then had quantities of vinegar poured down its throat, while the other dog received no treatment. Blackish burns appeared in the mouth of the dog that received no treatment but the acetic acid in the vinegar successfully neutralised the corrosive effects of nicotine in the other, and no signs of chemical injury appeared. Clearly the Count had learnt a lot about the chemistry of nicotine, and when Gustave put up a struggle, causing the nicotine to be splashed around, the Count did his best to conceal the evidence using vinegar.

The case went to trial. The Count and Countess did their best to accuse each other of the crime, but the evidence was damning. Somewhat inexplicably, Countess Lydie Bocarmé was found not guilty. Count Bocarmé was sentenced to death by guillotine.

Agatha and nicotine
In Agatha Christie's *Three Act Tragedy*, all three murders are committed by the ingestion of nicotine. The first takes place at the house of Sir Charles Carmichael, a stage actor. Sir Charles invites a group of friends and acquaintances to dinner at his house, including Hercule Poirot. Cocktails are handed round and the local vicar, the Reverend Babbington, sips at his before pulling a face, clearly indicating that he does not like the flavour. But being every inch the polite English vicar, he does not wish to offend his host or appear unsophisticated in his tastes, and he drinks down the whole glass. Within minutes Babbington's face is convulsed; he rises to his feet, sways a bit

and collapses. Two minutes later he is dead, long before a doctor can arrive to treat him.

The death is initially treated as suspicious because of Babbington's clear distaste for the cocktail. There must have been something in it, something so strongly flavoured that it could not be masked by gin and vermouth. Also, there was the sudden seizure he experienced shortly after drinking the cocktail. 'Seizure' is a vague term that doesn't really indicate any particular illness or poison. Such a general term can refer to any number of symptoms and causes. Only strychnine or tetanus produces characteristic seizures that could be used as a diagnosis. The cocktail glass Babbington drank from is taken for analysis, but tests reveal that it contains nothing but gin and vermouth. Wild theories are put forward about hypodermics of untraceable arrow poisons from South America being injected into the unfortunate clergyman. With no evidence of poison in his cocktail glass, and no idea of a motive for doing away with a rather dull vicar, the death is attributed to natural causes. But suspicion lingers, because although Babbington was advanced in years he was in good health.

Weeks later, there is another dinner party, this time hosted by Sir Charles's friend, Sir Bartholomew Strange, at his home in Yorkshire. Most of the guests had also been present at the previous party, when the Reverend Babbington died. After the meal port is served, and Sir Bartholomew is soon taken ill. He dies within minutes, showing similar symptoms to those of Babbington. Sir Bartholomew Strange was in the prime of his life, and a heart attack or stroke seems an unlikely cause of death. He was in good spirits on the night of his death, so suicide is also eliminated. A post-mortem is carried out on Sir Bartholomew's body, and his port glass is sent for analysis. No trace of poison is found in the glass of port, but the post-mortem reveals that Sir Bartholomew died of nicotine poisoning.

There may be very little physical evidence of nicotine poisoning post-mortem. As we have seen pure nicotine is corrosive, so with large doses there may be signs of burning in the mouth and throat. However, even if there are no physical signs, nicotine can still be identified, as it can be extracted from human tissue. The method eventually figured out by Jean Stas (see pages 169–170) remains little changed, even today, though modern scientists can use a variety of subtle chromatographic techniques to isolate and identify nicotine, and find out how much of it is present. Chromatography methods were still fairly basic when Agatha Christie was writing *Three Act Tragedy*, and pathologists would have had to rely on chemical reactions to detect the presence of nicotine. There was a variety of tests available, using compounds such as gold chloride (Au_2Cl_6) or picric acid ($C_6H_2(NO_2)_3OH$).[*] Nicotine reacts with these chemicals to produce crystals such as nicotine picrate (after reaction with picric acid). The most sensitive test in the 1920s used silicotungstic acid ($H_4[W_{12}SiO_{40}]$) and dilute hydrochloric acid, and could detect nicotine in quantities as tiny as one part in 300,000. If nicotine was present the reaction mixture clouded almost immediately, and crystals appeared when the mixture was left to stand. The crystals could be collected, washed and weighed to determine the amount of nicotine present in the sample.

Post-mortem results for nicotine may be complicated by the presence of nicotine from other sources. If the victim was a heavy smoker, as was the case with Sir Bartholomew, nicotine would already be present in the body. Nicotine is also rapidly metabolised into cotinine in the body, with a half-life of only one to two hours. Cotinine has a half-life of approximately twenty hours and is used as an indicator of exposure to tobacco, as it can still be detected days or even weeks later. Cotinine will also interact with nicotinic receptors but with a much lower potency than nicotine. Blood serum levels of nicotine or cotinine above 2mg/litre are associated with serious toxicity.

Even though Sir Bartholomew was a heavy smoker, nicotine poisoning is confirmed as the cause of death. It is virtually

---

[*]Picric acid is also known as trinitrophenol, which is chemically very similar to trinitrotoluene, or TNT. Consequently picric acid is also explosive.

impossible to smoke enough cigarettes or cigars in one go to be at risk from nicotine poisoning.* Nicotine is a fast-acting poison, so a large dose must have been administered to Sir Bartholomew very shortly before he died, but no one can explain how. Poirot suggests that a clear colourless liquid such as nicotine, added to the port Sir Bartholomew had drunk, might dilute its colour. However, the traditional volume of a port glass is 190ml, and an enormous quantity of nicotine, well above the lethal dose, could have been added to a glass of port before the colour was sufficiently diluted to notice. As Poirot points out, even if the colour had been diluted, the cut of the port glass would help hide the presence of nicotine.

In the light of events at Sir Bartholomew's dinner party, the death of the Reverend Babbington looks even more suspicious, and the vicar's body is exhumed. A post-mortem reveals the presence of nicotine, even though the Reverend wasn't a smoker. Any nicotine in his body from passive smoking would have been only in very low quantities, so it was easier in this case to attribute death to deliberate poisoning. Metabolic processes in the body stop at the point of death, and nicotine is remarkably stable with regard to decomposition post-mortem; it has been detected in human remains months after burial. The presence of nicotine in Babbington's remains, and whether there was enough for a fatal dose, should, therefore, have been easy to determine.

As the title of the book suggests, there had to be a third murder, and this time there was no doubt as to how the poison was administered. The final victim is Mrs de Rushbridger, a patient at the sanatorium that Sir Bartholomew ran; and her dose of nicotine is delivered in a box of liqueur chocolates. Only one chocolate is eaten and death occurs very rapidly, just two minutes later. There must have been a very large dose contained inside that single chocolate. The space inside a liqueur chocolate is quite small, so almost all of it must have

---

*There is, however, at least one historical example of death occurring during an ill-advised pipe-smoking contest.

been taken up with nicotine, leaving little space for liqueur to help disguise the flavour. Mrs de Rushbridger was either taken by surprise by the unusual flavour or was too polite to spit out the chocolate.

✧✦

There is no doubt about the cause of death in any of the three murders in *Three Act Tragedy,* but how easy would it have been for the murderer to obtain nicotine? There was a range of options open to the 1930s would-be nicotine-poisoner. One method would be grow-your-own, but the amount of nicotine in a plant varies with species and age, and this is perhaps the least reliable method. Another method would be to extract nicotine from tobacco products like cigars or cigarettes. The amount of tobacco per 1,000 cigarettes in 1960 was 1kg, giving an average of 1g tobacco per cigarette; as few as 35 cigarettes could be used to obtain a lethal oral dose. The amount of tobacco in cigarettes has dropped considerably since the 1930s because of the use of reconstituted tobacco and additives. However, since 1999 the average amount of *nicotine* has been increasing, year on year by approximately 1.3 per cent. The overall result is that today, you would need around 60 cigarettes to extract the same amount of nicotine. However, the easiest method in 1935 would have been to purify nicotine from a nicotine-based insecticide, and this was the method chosen by the murderer in *Three Act Tragedy.* In 1935 the concentration of nicotine in insecticides would have been very high, perhaps as much as 43 per cent.

There have been mutterings about the implausibility of the motive behind the three murders in *Three Act Tragedy.* For the American publication of the book, the motive was changed significantly. But, as usual for Christie, there is little to complain about in the science. From a practical point of view, the means and methods of murder are credible and accurate.

OPIUM

Sad Cypress

I do by no means deny that some truths have been delivered to the world in regard to opium: thus, it has been repeatedly affirmed by the learned that opium is a tawny brown in colour – and this, take notice, I grant; secondly, that it is rather dear, which I also grant – for, in my time, East Indian opium has been three guineas a pound, and Turkey eight; and, thirdly, that, if you eat a good deal of it, most probably you must do what is disagreeable to any man of regular habits – viz., die.

Thomas de Quincey, *Confessions of an English Opium Eater*

Opium has been with us for millennia. Poppies and their extracts are mentioned in the Ebers papyrus of 1500BC but descriptions of their effects on people have been found in Sumerian records dating back 6,000 years or more. Opium is perhaps the oldest medicine, and its astonishing efficacy means

it is still in use in various refined forms today.* It has brought
relief from pain and suffering to millions, as well as serving as
an inspiration for poets and painters, but it has also brought
about untold misery. Opium, its constituents and several of its
derivatives are highly addictive. The overpowering desire or
need for these drugs overrides every other thought, and the
demand for them has brought about crimes from petty thefts
to international wars.

In nineteenth-century England, opium, in the form of
laudanum, was a part of everyday life in the way that cigarettes,
alcohol and paracetamol are today. Opium could be purchased
at almost any pharmacy or grocer's, without question. Today,
despite strict laws and, in many places, severe punishments, it is
estimated that 9.2 million people across the world use one of
opium's most powerful and destructive derivatives – heroin.

Agatha Christie was well aware of the many sides of opium's
character. Opium and its derivatives are mentioned in more
than a dozen of her books; Christie had her characters relieved
from pain, sedated, addicted and murdered by this class of drug.
Two of the nine victims that Agatha killed using opium
compounds appear in the novel *Sad Cypress*, written in 1940,
and the plot has some similarities with real-life cases. Both
victims, Laura Welman and Mary Gerrard, are killed with
morphine, the most commonplace of the biologically active
compounds found in opium. Elinor Carlisle stands accused of
both murders, and the case against her appears to be conclusive.
That is until Hercule Poirot is called upon to use his little grey
cells and save her from the gallows.

---

*There are references to poppy plants, seeds, grains and stalks in
several places in the Ebers papyrus, a collection of Egyptian medical
knowledge. One application was for diseased toes. The poppy parts
were to be incorporated into a poultice and held there for four days.
Pods of the poppy plant were also recommended to stop the cries of
a child, as well as to relieve pain. The authors seemed well aware of
the poppy's analgesic and narcotic effects.

The opium story
Opium is the name given to the crude extract obtained from
poppy plants. There are many species of poppy, several of which
contain useful amounts of opium, but *Papaver somniferum* is the
species cultivated specifically for the opium it contains. This
species has been grown in the Middle East since at least 3400BC,
though the plant originated in Turkey. The poppy in its many
varieties now grows throughout the world, both wild and in
legal and illegal cultivations (for food, medicines and street
drugs), as well as in gardens as ornamental flowers. Opium for
illegal use is obtained from poppy plants by slicing the green
seed head with a razor blade, allowing a milky sap to ooze out.
After a day the dried sap will have turned brown, and the
gummy mass is scraped off, pressed into cakes of raw opium
and allowed to dry. For legal supplies to the pharmaceutical
industry, farmers first remove the seed pods (with the seeds
sold for use in bread and cakes). Morphine is then extracted
from the remaining stems and leaves, using solvents.

Throughout the centuries of its use, opium has been the
medication of choice for a huge array of pain-causing
conditions. The drug is ineffective in treating the underlying
causes of disease, but it is a powerful analgesic (pain-relieving
drug) that gives the patient a feeling of well-being; all things
considered, opium was, until relatively recently, one of the very
few drugs that would have been of any real benefit to a patient.
Many herbs, extracts, tinctures and products of the chemist's
laboratory were used as medicines, some of which were
innocuous, but at worst they were potentially lethal. Almost
any compound seen to produce an effect on an individual's
constitution was thought to be beneficial in some way, even if
the person initially appeared to get worse. Until science
determined the root cause of diseases, doctors were effectively
working blind, and effective treatments were discovered more
by luck than judgement. Medicine in the mid-nineteenth
century was neatly summed up by Oliver Wendell Holmes Jr,
in his address to the annual meeting of the Massachusetts
Medical Association in 1860: 'Throw out opium, which the

Creator himself seems to prescribe … and the vapours which produce the miracle of anaesthesia, and I firmly believe that if the whole *materia medica* as now used could be sunk to the bottom of the sea, it would be all the better for mankind – and all the worse for the fishes.'

In addition to its analgesic properties, opium in higher doses also acts as a sedative. An early form of anaesthesia was the *spongia somnifera,* or sleep-inducing sponge, which was used to lessen the pain and horror of medieval surgery. The sponge was soaked in a solution of water and wine, mixed with opium, lettuce, hemlock, hyoscyamus (henbane), mulberry juice, mandragora (mandrake) and ivy.* Once prepared, the sponge was dried out and stored until it was needed; it was then moistened so the juice could be drunk. The same sponge could be used a number of times.

Up until the middle of the nineteenth century, opium was mixed with a staggering variety of other ingredients to produce a range of remedies for almost every conceivable ailment. These other ingredients were usually inactive, sometimes lethal and occasionally very strange; combinations included powdered pearls, chloroform, wine, belladonna and amber. The mixture of opium with alcohol, now commonly referred to as laudanum, was discovered by Paracelsus (1493–1541), a Swiss-born alchemist, who realised opium was more soluble in alcohol than water, and that it was more effective in this form. Two depressant drugs taken together, such as alcohol and opium, enhance the effects of each other, making laudanum a potent pain-reliever, but also enhancing the depressant effects, thereby increasing the risk of coma and death. In the nineteenth century 'tincture of opium' consisted of the powdered drug dissolved in alcohol in a 10 per cent mixture, though in reality the concentrations of both opium and alcohol varied

---

*The *spongia somnifera* may well have relieved the pain of surgery, but with these plants and the lethal hemlock in the mix it seems a thoroughly risky way of doing it.

enormously. Its use became widespread and prices plummeted; tincture of opium became cheaper than gin or wine, and there were no restrictions on its sale. By around 1850, the annual consumption of opium averaged 5g per person. It was even spooned into the mouths of infants to relieve teething troubles. The result was that many people became addicted, including the poet Samuel Taylor Coleridge, Abraham Lincoln's wife, Mary, and even Queen Victoria herself.

Opium addiction
An addiction to opium through ingesting it is actually quite difficult to acquire. Opium with limited processing and purification represents a diluted form of its active components, and when eaten or drunk these compounds are metabolised too quickly to produce the euphoria experience by injection of the drug. So, large quantities of opium have to be consumed to produce an addiction, but when prices collapsed as they did in nineteenth-century Europe and North America many people were able to do just that. Those taking high doses may have experienced vivid, spectacular and sometimes terrifying dreams, which is perhaps why opium became popular with artists.

The isolation and increasing availability of the active principles within opium – morphine – combined with the invention of the hypodermic syringe (around 1848) led to a dramatic increase in addiction. When the drug is delivered directly into the bloodstream, the feelings of euphoria are more rapidly achieved than when it is ingested (by eating it), and there is very little time to metabolise the drug to less potent compounds. Despite increasing and occasionally very public cases of addiction, nothing was done to regulate the sale and use of opium in Britain until the 1868 Pharmacy Act. This Act limited the sale of opium and related products to licensed premises, but the laws were largely ignored until greater penalties and more serious enforcement were introduced in 1908.

Opium contains around 50 different alkaloids, only some of which are of pharmacological interest. These include morphine and codeine; noscapine,* which has cough-suppressant properties; and papavereine, a smooth muscle relaxant. Compounds extracted from poppies are classed as opiates, whereas those that produce opiate or morphine-like reactions in the body are called opioids. Opium is still part of the *British Pharmacopoeia* even today,† but the number of preparations has fallen from an all-time high of 38 in 1864 to just four in 1998. Opium prescriptions are now very rare; it is the isolated and purified opiates and opioids that are used in medicine.

Morphine

Of all the alkaloids found within the poppy plant, morphine is the best known, and it has powerful pain-relieving properties. German chemist Friedrich Sertürner (1783–1841) was the first to isolate it from opium. When Sertürner was just 16 years old, he heard doctors complaining that some samples of opium were more potent than others. He reasoned that opium was an impure mixture of compounds containing perhaps one active ingredient, and the quantity of this was likely to vary. It took him several years to isolate a white crystalline solid from raw opium, and he chose to test this in powder form on himself and three friends. They experienced severe nausea, and fell asleep for 24 hours. Sertürner therefore named the compound morphine, after Morpheus, the Greek god of dreams and son of Somnus, the god (or personification) of sleep who gives his name to the opium poppy's specific name, *somniferum*. Hardly anyone paid attention to this momentous discovery at the time. Later, when suffering from severe toothache, the pain of which even opium could not relieve, Sertürner tried his white powder again, but in a smaller dose this time. He remained awake, and his toothache disappeared completely. This time the medical profession paid attention. The benefits of morphine over opium were obvious.

---

*Formerly known as narcotine.
†Tincture of opium is also listed in the *United States Pharmacopeia*.

Codeine

More compounds have since been isolated from opium that have roles in modern medicine. Codeine, the second most abundant alkaloid in opium, was first isolated in 1832 by Pierre Jean Robiquet (1780–1840). Structurally, codeine is very similar to morphine, differing by only one methyl group ($-CH_3$). Inside the body, enzymes remove the methyl group and replace it with a hydrogen atom, effectively converting codeine into morphine. Codeine therefore has moderate pain-relieving properties, but it is much less addictive.[*] It is still used in cough syrups because it relaxes the muscles in the throat and dries up secretions. It is also used in diarrhoea remedies, because codeine slows the muscles controlling the gut (so the watery faeces cannot be so easily expelled).

Diamorphine

By the late nineteenth century the search was on for ever more potent pain relievers, and chemists in pharmaceutical laboratories tinkered with the structure of morphine to see if the drug could be improved. One successful modification was the addition of two acetyl groups ($CH_3C(O)-$) to form diacetylmorphine or diamorphine. The addition of the acetyl groups increases the solubility of the molecule in fats, and enables it to cross the blood–brain barrier more easily than morphine can. Once inside the brain, enzymes quickly remove the acetyl groups, converting the diamorphine back into morphine, where it can interact directly with opioid receptors in the brain. Consequently diamorphine acts more quickly than morphine, and it is therefore far more potent. It makes you feel like a hero, so they called it heroin. Diamorphine was first made in the laboratories of the German chemical company Bayer. The chemical process to convert morphine to heroin was simple, allowing huge quantities of this powerful drug to

[*]The conversion of codeine to morphine is slow. Codeine is therefore far less addictive than morphine or heroin (and is more moderate in its pain relief), but it can still be addictive if used regularly.

be produced cheaply and easily. It was released onto the market in 1898.

Initially heroin was touted as an effective and non-addictive form of morphine. It was recommended for adults and children suffering any form of pain or discomfort. Heroin is more effective at suppressing coughs, and causes less constipation in its users than morphine. It seemed an ideal drug. However, the intensity of the high and the rapidity with which it can be achieved compared to morphine makes it far more addictive. The withdrawal symptoms from the drug are also more intense than with morphine, increasing the desire for another hit. Four years after heroin's release onto the market its addictive properties had been realised, and many countries had banned it. Most countries around the world still ban the import, production and sale of heroin, as the risk of addiction is thought to far outweigh the benefits of the drug. Britain is a notable exception; it continues to prescribe the drug in palliative care, but under very strict rules and regulations. When prescribed medically the name diamorphine (or diacetylmorphine) is used; 'heroin' is the name usually reserved for the drug when it is manufactured, transported and sold illegally. The ease of growing poppies, extracting opium, refining the morphine it contains and modifying this into heroin make it an extremely profitable street drug (the main difficulty being that you need an awful lot of raw material). One kilogram of raw opium can produce 100g of morphine or heroin[*]. It would take an annual harvest of at least 10,000 poppy plants to supply a typical heroin user for a year.

How opiates work
Once absorbed into the body, the effects of morphine, codeine and heroin are similar, as enzymes inside the body quickly

---

[*]The amount in a 'wrap' of drugs on the street varies enormously; street drugs are almost never pure. A typical batch of heroin bought on the street can be anywhere between 10 and 40 per cent heroin; the rest is by-products from the manufacturing process, or other substances deliberately added to dilute the heroin and increase the profit.

convert codeine and heroin into morphine. Heroin is inactive in the body, and must be converted into morphine and other compounds to produce an effect. Direct injection into the bloodstream is the most effective way of delivering these drugs and, using this method of administration, their effects are observable almost instantly. Ingested morphine, codeine and heroin are absorbed into the bloodstream from all parts of the gastrointestinal tract, except the mouth. The liver is the main organ of metabolism; most ingested heroin is converted into morphine in this organ within two or three minutes.

Morphine mimics endogenous (meaning 'synthesised by the body') opioids such as endorphins (or 'endogenous morphine'). Structurally, endorphins are very different from morphine, but the results in the body are similar. Endorphins and morphine both activate opioid receptors, inducing feelings of pain relief, sleep, pleasure and relaxation.

Opioid receptors are found in the nerve cells of the brain, spinal cord and gut. There are four main types; δ (delta), μ (mu), κ (kappa) and nociception (NOP) receptors, and there are further subtypes within these groups. Interaction of morphine with μ receptors leads to pain relief, but also euphoria and slow, shallow breathing. This interaction seems to be largely responsible for addicts' dependence on morphine. The δ and κ receptor interactions also produce some pain relief, and binding with κ may be responsible for some sedative effects. NOP receptors appear to be involved in the regulation of several brain functions, instinctive and emotional responses in particular.

From a medical point of view, the most important effect produced by morphine is pain relief. Morphine is still considered to be one of the best pain-relieving medications available, and it is the gold standard to which new drugs are compared. The interaction of morphine with opioid receptors in the cerebral cortex, the higher functioning part of the brain, modifies our *perception* of pain. A person under the influence of morphine may continue to be aware of pain, but is no longer concerned about it. Morphine is most effective when given as a subcutaneous or intravenous injection, in a dose of 10mg for 70kg of body weight; oral administration of morphine is only one-sixth as effective. Morphine is poorly soluble in water and is usually administered as a salt, often the hydrochloride or sulfate; these are colourless and odourless, but have a bitter taste.

The breakdown of morphine in the body produces compounds known as glucuronide conjugates. These are water-soluble and can be excreted in the urine or in bile. Some of these compounds also have pain-relieving properties. For example, a trial on 20 cancer patients reported that treatment with morphine-6-glucuronide gave pain relief without sedation or euphoria.

The effects of morphine generally last between three and six hours, but this varies between individuals. This relatively short time-span often means that further doses may be required to manage severe pain. After the first few doses the body begins to become accustomed to the presence of morphine, and ever-higher doses are required for effective pain relief. After daily treatment of a few weeks, a patient may need 100 times more morphine than at the start. The body adapts to the pain relief quicker than to the level of euphoria, so increasing the dosage to manage pain also increases the feelings of euphoria. This is the development of tolerance. Opioid receptors are normally stimulated by chemicals produced within the body. For the body to still respond to these endogenous compounds in the presence of morphine, more receptors must be produced. As the number of receptors rises, the same amount of morphine will no longer produce the same effect, so the dose must be increased; an ever-increasing amount of morphine is required to achieve the same initial levels of relief. The longer treatment is continued, the more likely it is that an addiction will form.

The euphoric and addictive properties of morphine are ultimately caused by the drug's effect on dopamine levels in the brain. Opioid receptors are indirectly involved in the release of dopamine, a chemical within the body that plays an important role in feelings of well-being. The brain releases dopamine to reinforce behaviours that are important for survival and persistence of the species – eating food, for example, and sexual contact. The interaction of morphine with opioid receptors triggers other receptors to release more dopamine.

When the body has become accustomed to the presence of morphine, a sudden decrease or absence of the drug can produce withdrawal symptoms. Cells suddenly find that there are a large number of opioid receptors with little or no morphine to stimulate them. This can lead to a wide range of symptoms that include anxiety, sweating, vomiting, diarrhoea, chills (leading to goose bumps, and the origin of the phrase 'going cold turkey'), bone pain, heart arrhythmias, depression and headaches. These symptoms, though very unpleasant and unquestionably deeply distressing, are rarely fatal. With prolonged abstinence the number of opioid receptors decreases, and after a number of weeks or months they can return to normal. An addict returning to heroin or morphine use after a prolonged absence can easily overdose by using their 'normal' hit, because they have lost their tolerance to the drug. However, it should be said that addiction rarely occurs in medical users of morphine and related products. For example, 90 per cent of American soldiers who used heroin regularly in Vietnam did not use it again after returning home.

The interaction of morphine with opioid receptors in other parts of the body produces some of the side effects experienced by its users. Interaction with opioid receptors in the gut reduces the activity of the muscles that move food through the intestines. The result is often constipation, but a more serious consequence can be a delay in the release of drugs because of their retention in the stomach for hours longer than usual. Other common side effects include pinpoint pupils, nausea, vomiting, itching, fainting upon standing up, mental 'clouding' and retention of urine. Some side effects of opiates and opioids can be advantageous. Morphine-like drugs, such as codeine, are still used in cough medicines, to suppress the cough reflex, and to treat irritable bowel syndrome. But in some people the side effects can be more serious. Some individuals are allergic to morphine, and in rare cases morphine may cause mania, delirium and, rarely in adults but more commonly in infants, convulsions.

One very serious side effect of morphine, and therefore of heroin and codeine, is a reduction in breathing rate. This is what kills in cases of overdose or poisoning. Usually, breathing rates are carefully monitored and controlled in the body. Receptors in the body are sensitive to the amount of carbon dioxide being expelled as well as to low levels of oxygen (hypoxia). During periods of increased metabolic rate, higher levels of carbon dioxide are produced than usual. The excess carbon dioxide is converted by enzymes into carbonic acid, which lowers the pH of the blood. The body normally responds by increasing the rate and intensity of breathing to expel excess carbon dioxide and to increase the amount of oxygen entering the lungs. Opioids reduce the sensitivity of the respiratory centre (a region of the brain that controls the muscles responsible for inhalation and exhalation) to carbon dioxide, and depress the automatic activity of this region so breathing becomes slower, and may stop altogether during sleep. The lethal dose of morphine is generally taken to be between 100 and 300mg, though addicts can tolerate 10 to 20 times as much.

The symptoms of morphine poisoning (which are very similar to those of heroin or other opiates) appear within five or ten minutes after injection, or 15 to 40 minutes after ingestion. Sedation deepens rapidly into coma; the pupils contract to pinpoints and there is a substantial reduction in the rate of respiration. Death occurs from respiratory failure. By contrast, excitement and convulsions are not uncommon in codeine or heroin poisonings.

Is there an antidote?

A number of drugs have been developed to act as specific antidotes for a morphine overdose, by displacing it from the opioid receptors. Perhaps the most successful of these drugs is naloxone, a compound developed in the 1960s, which is a pure opioid antagonist. Naloxone[*] is structurally very similar to morphine, but subtle differences produce dramatically different results. The drug is administered by injection in cases of an opioid overdose, to restore normal breathing. When naloxone takes the place of morphine at opioid receptors, it does not trigger them. The effects of opioids are therefore mostly reversed, including suppression of breathing and pain relief, and this can occur within a few minutes.

Some real-life cases
The ease of obtaining morphine before 1920, and the fact that it was undetectable in cadavers until 1850, together with publicity from several high-profile cases undoubtedly means murderers have gone unpunished for their crimes in the past. Morphine and its derivatives have continued to be used as poisons in murder, even when regulations have attempted to control their distribution and forensic methods have improved to enable detection of morphine poisoning several years after the event.

Some years after *Sad Cypress*, Agatha Christie returned to morphine as a means of murder in her 1968 novel *By the Pricking of My Thumbs*, in which a series of murders occurs in a nursing home, with no apparent motive. A real-life case from 1935 may well have provided inspiration to the author.

Dorothea Waddingham was not a qualified nurse, but she was always referred to as Nurse Waddingham. She ran a care home in Nottingham with her husband; in 1935 there were three residents in the home, Louisa and Ada Baguley (a mother and daughter), and a Mrs Kemp, whose condition caused her a lot of pain and for which she was prescribed morphine. When Mrs Kemp died, Nurse Waddingham still had a considerable amount of morphine in her possession. The Baguleys, the two remaining residents, were persuaded to sign

[*]Marketed under the names Narcan, Nalone and Narcanti.

over all their money to Nurse Waddingham in return for being looked after in the home for the rest of their lives. This turned out to be a very short period of time.

At no point did the Baguleys complain about their treatment in the home, and the pair seemed to be very happy there. However, they required a considerable amount of care in return for the relatively low rent (£3 per week[*]) they were paying. The 50-year-old daughter, Ada, suffered from a degenerative disease, and her elderly mother was no longer able to care for her. Nurse Waddingham commented that 'they would have to pay five guineas a week each for no better treatment in hospital, and that is really the proper place for them'.[†]

Ada Baguley changed her will on 4 May 1935. Eight days later her mother Louisa was dead. No suspicion was attached to the death, and Ada continued to live at the home. On 10 September a friend of Ada's, Mrs Briggs, visited her at the home and found Ada in good spirits. She invited Ada to visit her on the following Thursday, and everyone agreed to the arrangement. But the next morning Ada was found in a coma, and a doctor was called. She had been unconscious since 2 a.m., but the doctor did not arrive until midday, three hours after he had been called. By the time he arrived Ada was dead.

The doctor was not surprised at Ada's death, but he had not expected it to come so soon. He examined the body, certified that death was due to cardiovascular degeneration, and issued a death certificate. Ada Baguley's cremation was arranged for 13 September, but it did not take place. A certain level of scrutiny is applied to cremation in Britain, in case someone is attempting to destroy evidence of foul play. In 1935 a second death certificate had to be issued by another doctor, and all cremations were reviewed before going ahead. In most cases, of course, this was purely a formality. However, the Cremation Referee in Nottingham at the time, Dr Cyril Banks, also happened to be the Medical Officer of Health. He knew Nurse

---

[*]Equivalent to around £180 (or $270) per week now.
[†]From *Poisoner in the Dock* by John Rowland.

Waddingham's home was not a registered nursing home, and no state registered nurse was on its staff. Banks decided to investigate Ada's death more fully, and ordered a post-mortem.

There was no indication of a natural cause of death. Ada's condition, disseminated sclerosis or 'creeping paralysis', was progressive but not at a stage where death might be expected. Analysts were called in to see if any poison had been administered. Considerable quantities of morphine were found in the stomach, spleen, kidneys and liver. The stomach contained 2.5 grains (approximately 150mg), and the body as a whole contained more than three grains. Given the rapid metabolism of morphine in the body, the actual dose was probably much higher. In the light of these results Ada's mother Louisa Baguley's body was swiftly exhumed; it too was found to contain considerable quantities of morphine-related compounds. Nurse Waddingham was found guilty of murder, and was hanged in April 1936.

No chapter about morphine can fail to mention the most notorious real-life morphine poisoner of all, Harold Shipman. Shipman was a medical doctor who worked as a general practitioner in the north of England. Britain's most prolific serial killer, he was found guilty in January 2000 of the murder of 15 people. A subsequent enquiry investigated the deaths of more than 1,000 of Shipman's patients, and he was found responsible for between 220 and 240 of them. The true total will probably never be known, because in many of the earlier cases there was too little evidence on which to determine what really happened.

The deaths occurred over a period of 23 years, but suspicion was only aroused because Shipman attempted to forge the will of his last victim, Kathleen Grundy, in 1998. His victims were usually elderly women living alone, though he also killed some men. They were usually healthy for their age, and with no apparent wish to die. His oldest victim was 93, Ann Cooper, and the youngest was 41-year-old Peter Lewis, who was terminally ill and whose death Shipman hastened. Most of his victims were killed with an injection of an opiate, usually diamorphine, though he sometimes used a large dose of sedatives.

Shipman was able to obtain large quantities of controlled drugs by a number of different methods. For example, in 1996 he prescribed and obtained as much as 12,000mg of diamorphine on a single occasion, in the name of a dying patient. That alone would have been sufficient to kill about 360 people. The enquiry into Shipman's murders led to significant changes in the regulation and control of certain prescription drugs such as diamorphine and morphine.

Much of the evidence against Shipman came from the examination of death certificates and medical records, as well as interviews with relatives of the deceased and work colleagues. In addition to this, nine exhumations were carried out, and post-mortem examination revealed the presence of substantial quantities of morphine. Many of Shipman's victims were cremated, and other deaths had occurred too long ago to obtain forensic evidence; after burial for more than four years morphine levels in a body cannot be reliably determined. There was no attempt to determine whether the drug used was heroin or morphine, or the quantity or method of administration. Analysis was conducted on the thigh muscle and liver where possible. Shipman had claimed many of his victims had a drug habit, but analysis of hair revealed only very low levels of morphine, consistent with individuals who had used over-the-counter medications containing codeine.

Shipman hanged himself in prison without ever revealing the total number of his victims, or his motives for killing them. In a will, dated 1979, Shipman had left everything to his wife. He had ticked the cremation box.

Agatha and morphine
The plot of *Sad Cypress* centres around two deaths caused by morphine poisoning. The prime suspect is Elinor Carlisle, a young woman hoping to marry Roddy Welman. The only

cloud on the couple's horizon is a lack of funds, but this is only a temporary set-back. The pair are both related to the wealthy Mrs Laura Welman and, as her closest relatives, they expect to inherit. Mrs Welman has recently suffered a stroke and is in poor health, so it is only a matter of time before the couple should be able to set a date for their big day. That is until Elinor receives an anonymous letter suggesting that Mary Gerrard, a pretty young companion of Mrs Welman, has been sucking up to the old lady, and the inheritance may be in jeopardy. The pair decide to visit Mrs Welman, ostensibly to see a much-loved relative, but also to protect their interests. The visit is going well until Roddy bumps into Mary Gerrard, whom he has not seen since childhood, and is bowled over by her beauty. Elinor is suddenly in danger of losing both her inheritance and her man to the same woman.

A week after the visit a telegram summons Elinor and Roddy back to Mrs Welman's house. The old lady had suffered a second stroke and was very ill. Elinor arrives at Mrs Welman's bedside in time to hear her request for a solicitor. Mrs Welman dies the following night, before the solicitor arrives. The death certificate is made out, with the death attributed to natural causes. No one is surprised at Mrs Welman's passing, but it is thought to be a little sooner than expected. The only surprise is that Mrs Welman died intestate. With no will, Mrs Welman's considerable inheritance goes to Elinor Carlisle as her closest living relative (Roddy was only related by marriage). Elinor and Roddy's marriage is called off, perhaps due to Roddy being uncomfortable at being financially reliant on his future wife, but maybe it has more to do with his infatuation with Mary Gerrard.

One month later, Elinor returns to the house to pack up her aunt's belongings. She invites Mary Gerrard and Mrs Welman's former nurse, Nurse Hopkins, to have lunch with her at the house. A plate of fish-paste sandwiches and tea is served. Nurse Hopkins then offers to help Elinor turn out Mrs Welman's clothes. While the two are busy in another room Mary Gerrard slumps down in her seat and falls asleep. An hour later Mary

cannot be woken, and is clearly very ill. A doctor is called but
Mary dies shortly after he arrives.

A post-mortem examination of Mary's body reveals that she
died from a rare form of morphine poisoning – foudroyante,
from the French meaning 'violent'. There are three forms of
opium poisoning, as described in Alexander Blyth's book
Poisons, Their Effects and Detection, which Agatha Christie is
likely to have read. Morphine poisoning usually causes a period
of excitement, followed by narcosis and coma with the
symptoms appearing between 30 minutes and one hour after
exposure. The foudroyante form proceeds very rapidly, yielding
deep sleep within five or ten minutes and death within a few
hours; the pupils remain dilated, unlike the pinpoint pupils
usually seen in opioid users. A third, exceptionally rare, form of
morphine poisoning leads to convulsions, but no coma.

Mary has clearly been deliberately poisoned, and Elinor
immediately comes under suspicion. She had made an
unfortunate comment the day Mary died. When she purchased
the fish paste for the sandwiches from the village shop she said,
'one used to be rather afraid of eating fish pastes. There have
been cases of ptomaine poisoning* from them, haven't there?'
Despite her apparent concerns Elinor purchased two jars of fish
paste. All three women at lunch had eaten fish-paste sandwiches,
and all three would have been expected to be prostrate with
vomiting and diarrhoea if they had suffered 'ptomaine poisoning'.
As Poirot points out, if the intention had been to imply Mary
Gerrard had died of this, the choice of morphine was a poor
one. Other poisons could have imitated the symptoms much
more closely. Poirot even suggests that the poisoner should have
chosen something like atropine, if that had been their plan.

It would have been simple to determine morphine poisoning
from the post-mortem examination of Mary Gerrard's body.

---

*Ptomaine poisoning is an old-fashioned term for food poisoning.

The easy availability of morphine up until the early twentieth century meant that it was not an uncommon murder method, and scientists had to work hard to develop chemical tests to identify poisons in cases of suspicious deaths.

The real-life Buchanan case brought to light the problems of distinguishing between the different alkaloids that might be present in a body. In 1892, Dr Robert Buchanan was living in New York after divorcing his first wife, and he had taken up with Anna Sutherland, the madam of a brothel. Anna had amassed a huge fortune through her business, and Buchanan decided to marry her. Her fortune was clearly not enough for Buchanan, though, and he insured Anna's life for $50,000. When Anna died of a cerebral haemorrhage Buchanan was quick to collect the insurance money. He hurried back to his native Nova Scotia and remarried his first wife, just three weeks after Anna's death.

Buchanan had almost got away with murder, but friends of Anna were suspicious and thought Buchanan had poisoned her. Two years earlier, Buchanan had taken a particular interest in the case of Carlyle Harris, who had murdered his wife using an overdose of morphine. The authorities had been alerted to Carlyle's use of the poison by the appearance of Mrs Harris's eyes after death. The morphine had caused the pupils to contract to pinpoints; Carlyle Harris was subsequently found guilty of murder. Buchanan called Harris a 'bungling fool' and a 'stupid amateur', and pointed out to his friends that if Harris had used atropine eyedrops they would have counteracted the effect of morphine, and no one would have been suspicious. A nurse who attended Anna Sutherland during her final illness noticed Buchanan doing just that – putting drops into Anna's eyes when there was no obvious need to do so.

Anna's body was exhumed, and a post-mortem determined that death was due to a lethal dose of morphine, but the jury needed to be convinced. To demonstrate to the jury the effect of atropine and morphine a cat was brought into the courtroom. Both drugs were administered to the cat to demonstrate the effects on its eyes. This was a straightforward demonstration when compared to the difficulties in proving that sufficient morphine had been administered to Anna to kill her.

At the time there were a number of chemical reactions that could be carried out on suspect materials, and characteristic changes in colour would identify the presence of certain compounds. Buchanan's defence made a great show of the unreliability of these chemical colour tests. The best-known test for morphine at the time was the Pellagri test; the suspect substance was dissolved in concentrated hydrochloric acid, and a few drops of concentrated sulfuric acid were then added. Next, the resulting mixture was evaporated. A glowing red colour in the residue indicated the presence of morphine. By adding dilute hydrochloric acid, sodium carbonate and tincture of iodine (iodine dissolved in water and alcohol) to the mixture, the glowing red was transformed into green.

The Pellagri test, and many others, had been carefully carried out by Rudolph August Witthaus (1846–1915), a celebrated forensic chemist, in the Buchanan case. However, the defence produced another expert witness, Victor C. Vaughan, professor of chemistry at the University of Michigan. Vaughan claimed that cadaveric alkaloids (alkaloids produced in animal bodies by the decay process) gave the same results as morphine. A series of chemical tests were performed in the court that claimed to show positive colour-test results from extracts obtained from a decayed dog's pancreas. A courtroom was not the ideal place to carry out complex chemical tests, and Vaughan skipped a few of the steps he claimed were not important to the final outcome. Test after test was carried out on test tubes containing morphine, and others containing cadaveric alkaloids. The colours Vaughan produced in his test tubes did not always match up to those stated in the textbooks; either way, the jury became baffled by the array of colours and descriptions of chemical processes. They were left with the impression that, even if the colours produced did not match the textbook descriptions, the same colours were produced by both morphine and cadaveric alkaloids, and the two compounds were essentially indistinguishable. Buchanan was convicted of the murder (and subsequently went to the electric chair), but this was based on *other evidence* brought against him. The scientific evidence appeared discredited, and newspapers rushed to publish the findings – the reliable Pellagri test was not so reliable after all.

Though the tests performed by Vaughan for the jury did not stand up to the high level of standard procedures expected for such a serious crime as murder, the public's confidence in forensic science was badly shaken. Witthaus had pointed out during the Buchanan trial the importance of performing several tests and comparing the results. Several compounds may give the same result in a few chemical tests, but not all. When preparing for the trial, Witthaus had carried out every known test for the presence of morphine, as well as carrying out physiological studies on frogs. Anna Sutherland had certainly died of morphine poisoning, but a simple and reliable test needed to be found that would convince a jury. After the trial, to reassert the importance and reliability of forensic science, tremendous effort went into establishing reliable, reproducible and unmistakable tests for poisonous compounds. A new level of rigour was brought to the science of forensics, and these processes continue to be tested, improved and replaced by ever more robust methods to this day.

By 1955 there were 30 different methods for testing for morphine. As well as colour tests, pure compounds isolated from cadavers could be identified by their melting points and the shapes of crystals formed when the compounds were converted into salts. Physiological experiments could also be carried out to confirm the effects of a poison in cases where no specific chemical test was available. Even by 1940 (the year *Sad Cypress* was published), the level of scrutiny applied to the analysis and identification of substances in suspected poisoning cases would have been high.

In *Sad Cypress*, Mary Gerrard's murder means the circumstances surrounding Mrs Welman's death also look suspicious, and an exhumation is ordered. By this time Mrs Welman has been buried for more than a month, but a post-mortem examination is able to confirm that her death was due to morphine poisoning. Although chemical compounds are metabolised in a living body, these processes halt when the body dies; but this does not mean that the drugs and their metabolites are conveniently preserved, frozen in position, waiting for a pathologist to come along and reconstruct the crime by their analysis. A whole new series of chemical reactions can occur during decomposition that may convert any drugs present into new compounds, though many are remarkably resistant to this process. The final metabolite of morphine, the glucuronide conjugates (see page 184), can be slowly converted back to morphine after death. Unless precautions are taken, chemical reactions, even in sample vessels, can continue to degrade morphine products after their removal from the body. Also, the longer a body has been buried, the more water is lost from the tissues, and this can concentrate the amount of any drug present, so the extent of desiccation must be taken into account.

The most common samples to be analysed for drugs in post-mortem examinations are from the liver, blood and urine. But, depending on the state of decay, not all of these may be available. In Mrs Welman's case liver samples might have been the best option; there would be no urine to analyse, and if she was embalmed (which is likely, as she was rich) her blood would have been significantly altered. Other tissues can be used for drug analysis in post-mortems, but they are much less reliable. For example, muscle tissue may contain the highest drug load from the body, but there are difficulties in extracting these drugs, and they may not be evenly distributed throughout the muscle, which can potentially lead to huge errors. Analysis of hair is also used to determine drug use over time, and though hair is resistant to decomposition it also grows slowly, and drugs take time to find their way to this part of the body. If Mrs Welman had received only a single (massive) dose of morphine just before her death it would not have showed up in her hair. However, morphine can even be detected in maggots that have fed on human corpses poisoned by the drug.

Even if the quantity of morphine present is accurately determined, it can be difficult to say for sure whether this was the cause of death. Owing to people developing tolerance to opioids, determining levels in the body that might constitute a fatal dose is nearly impossible unless the patient's medical and recreational use of these drugs is well known. Codeine from over-the-counter prescriptions will also potentially show up in tests (as the compound is converted into morphine in the body). Even a strong liking for foods such as poppy-seed cake can skew the results of tests for opioids. Opiates are present in poppy seeds, but the amounts vary greatly. Eating poppy seeds is very unlikely to produce a high unless you eat an awful lot them (about 75g, I am told), but people eating food containing poppy seeds have been known to fail drug tests.

Mrs Welman's drug history would have been well known because her recent illness had required regular visits from the doctor, and two nurses were in constant attendance. Even if she was a dedicated eater of poppy-seed cake, there would only have been very small amounts of morphine present in her body. A pathologist would have been confident in asserting that her death had been due to morphine poisoning.

The evidence against Elinor Carlisle is building; when she stands trial it seems a certainty that she will be found guilty of murder. Poirot's expertise is employed in an attempt to get Elinor off the charges. In his usual methodical way, he sorts through the evidence, talks to those involved, dismisses red herrings and, finally, arrives at the truth.

The doctor attending Mrs Welman privately believes this might be a case of suicide or assisted suicide. The subject had been broached with the doctor by Mrs Welman herself, but the doctor had declined to give her the requested extra dose of medication. Despite being bedridden, Mrs Welman was an extremely resourceful woman, and the doctor never questioned that she could somehow have managed to obtain morphine, and take a lethal dose, if she had a mind to do so.

The Dangerous Drugs Act of 1920 ensured that opium, morphine and several other drugs could be obtained only with a doctor's prescription, and only from a registered pharmacy. Chemists, dispensers, doctors and nurses may have had access to morphine, but the quantities used were monitored, and stock checks were carried out to ensure that nothing was missing. The difficulty of obtaining morphine is discussed at some length in Agatha Christie's 1955 novel *Hickory Dickory Dock*. One of the characters in the novel is challenged to obtain three lethal poisons, one of which is morphine tartrate. He achieves the feat by posing as a doctor and stealing the drug from the poisons cupboard in a hospital dispensary, but other methods are also proposed. Any present-day poisoner wishing to use some of the methods suggested by Christie will be disappointed to discover that even these underhand methods are unlikely to be successful, as increased checks and controls have since been put in place.

In *Sad Cypress*, Christie describes a plausible situation that gives all of her chief suspects access to morphine. Nurse Hopkins had a tube of 20 half-grain morphine hydrochloride tablets in her bag, for use in treating a patient in the village who was suffering from carcinoma (skin cancer). The day before Mrs Welman's death, Nurse Hopkins found that she had lost the tube of tablets. Although she should have reported the loss, she did not do so at the time, as she thought she must have simply misplaced them. Later it was realised that the nurse's case had been left in the hall of Mrs Welman's house overnight, and anyone could have taken the morphine. The tube contained enough of the drug to kill several people not regularly using morphine or morphine-related products. A few of the tablets could have been dissolved and injected, if the murderer had access to a syringe, or they could have been added to Mrs Welman's food or drink. In her incapacitated state she was unlikely to be able to protest if what she was consuming tasted unusually bitter. Therefore the list of possible suspects extends beyond those in the medical profession.

Compared to Mrs Welman's death, Mary Gerrard's murder seems clear-cut. The overdose of morphine had been consumed at lunch, and only three people were in the room at the time. The set-up mimics a real-life case, which Agatha Christie may have used as a source of inspiration. In 1930 Sarah Hearn was accused of putting arsenic in some salmon sandwiches she had prepared for herself and her friends, William and Annie Thomas. The couple became ill some time after eating the sandwiches, and Annie was particularly unwell. While they were recovering at home, Sarah Hearn visited them to prepare a meal for the invalid and her husband. Two days later, Annie Thomas was dead. Analysis of her body showed the presence of large quantities of arsenic. Meanwhile Sarah Hearn had disappeared, leaving behind a note that hinted at suicide. In fact she had changed her name, and taken a housekeeping job in a neighbouring county. When her trial eventually came to court, Hearn's defence barrister discredited the scientific evidence by showing that large quantities of arsenic were present in the soil where the bodies had been buried, and this could easily have contaminated the samples analysed at post-mortem. The jury acquitted Sarah Hearn, but many questions remained unanswered. For example, how could she have ensured that she did not eat a poisoned sandwich along with her friends? And how could Hearn have added arsenic (which was thought to be in the form of weedkiller, and therefore stained blue) to the sandwiches without the colour showing in the bread? If Annie Thomas *was* murdered, her killer was never brought to justice.

Explaining how a poisoner can kill only one person at a lunch where three people ate the same food without revealing the murderer would result in some very convoluted descriptions. For the sake of clarity, this section contains a few major spoilers. If you do not want to know the result, look away now (or go to page 203).

In *Sad Cypress* Poirot realises that, though Elinor had prepared the sandwiches, they had been left unattended in the kitchen for some time and anyone from outside could have tampered with them. If an outside agent had poisoned the sandwiches it seems likely that all three people eating lunch would have been taken ill. If this unknown murderer had poisoned only one of the sandwiches it was pure bad luck that Mary Gerrard chose to eat it. Poirot dismisses this theory, and decides that the murderer must have been one of the three at the lunch. The analysis of Mary's stomach contents cannot discern whether the morphine was in the sandwiches or in the tea that she had also consumed. Elinor had not drunk any of the tea, but Nurse Hopkins had. If the morphine had been in the tea then it should also have poisoned Nurse Hopkins.

There was one telling clue that the police appeared to have overlooked, or they had not appreciated the significance of it. This was a pin-prick mark on Nurse Hopkins's wrist. Elinor notices it when the pair are washing up after lunch. Nurse Hopkins claims she got the mark from a thorn on a rose bush, but Poirot's meticulous attention to detail finds that the rose bush in question is a thornless variety. The mark could have been made by the needle of a hypodermic syringe, and this ties up nicely with another clue the murderer unwittingly left on the kitchen floor. This was a tiny fragment of a label from a pharmaceutical prescription.

The police claim that the label fragment came from a tube of morphine hydrochloride tablets, the tablets that had gone missing from Nurse Hopkins's bag, but closer inspection shows that this is not the case. In fact the label was from a tube

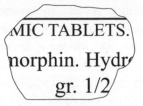

Fragment of label found on the kitchen floor in Sad Cypress.

of Apomorphine. The portion of the letter 'm' visible on the fragment of label shows it was lower case, and therefore could not have been from a label of Morphine hydrochloride tablets.[*]

The name apomorphine is slightly misleading, as it is structurally quite different from morphine. Apomorphine was originally synthesised by the method described by Agatha Christie in *Sad Cypress*: 'Apomorphine hydrochloride is a derivative of morphine, prepared by saponifying morphine by heating it with dilute hydrochloric acid in sealed tubes. The morphine loses one molecule of water.' Christie is correct in these statements, but the molecules of morphine subjected to this process do more than lose two atoms of hydrogen and one of oxygen. The atoms in the molecule actually undergo a major rearrangement, resulting in a compound significantly different from morphine, though it still binds to opioid receptors. The side effects of the compound mean it is unlikely to lead to an addiction, but people using it sometimes fall asleep. In the past, apomorphine has been used in aversion therapy for the treatment of anxiety, alcoholism, homosexuality and drug addiction.[*] The author William S. Burroughs felt it was the most effective treatment for opium addiction as it greatly

---

[*]Although the possibility that the label read Diamorphine does not seem to have been considered; diamorphine could also have caused the death of Mary Gerrard, and the toxicological analysis would be similar because of the swift metabolism of heroin to morphine in the body.

reduced the symptoms of withdrawal without being addictive. It was the only treatment, of the many that he tried, that gave him a lasting recovery from opiate addiction. However, outside of Burroughs's assertions, there is no evidence that apomorphine is a safe or effective treatment for opioid addiction.

Apomorphine is not a specific antidote to opioid poisoning as it can only remove any unabsorbed poison from the stomach. Administering apomorphine after the drug has been absorbed, or if it has been injected, would be of little benefit to the patient.

Nurse Hopkins added morphine to the pot of tea she prepared at lunch. Surprisingly, no one comments on any bitter taste to the tea – perhaps they were too polite to say anything. Mary Gerrard drinks some of the tea, as does Nurse Hopkins, but, by injecting herself with apomorphine shortly afterwards, Nurse Hopkins vomits violently to remove the poison before it can be absorbed into her body. Then, to ensure that Mary Gerrard succumbs to the effects of the morphine, Nurse Hopkins suggests helping Elinor turn out the clothes in a bedroom upstairs. Mary is left alone for an hour, so by the time her poisoning is 'discovered' it is too late to save her.

*Aversion therapy seeks to make the brain associate certain behaviours or actions with unpleasant stimuli, in an effort to wean the patient off them. For example, in alcohol aversion therapy, alcohol is administered with an emetic, causing vomiting. The individual associates alcohol with a deeply unpleasant experience; thousands of alcoholics have been treated since the 1930s, and there have been claims of significant success. Aversion therapy for homosexuals was mainly carried out in the 1960s, when homosexuality was still a crime in the UK; many people were referred for this through the courts. This involved injections of apomorphine to induce nausea and vomiting; images of men in various states of undress would then have been shown. Unsurprisingly, the treatment was unsuccessful. It also required hospitalisation, so the individual could be monitored and treated if dehydration occurred. There were doubtless deaths caused by the side effects of apomorphine treatment.

Dumb Witness

After some experiments made one day at my house upon the phosphorus, a little piece of it being left negligently upon the table in my chamber, the maid making the bed took it up in the bedclothes she had put on the table, not seeing the little piece. The person who lay afterwards in the bed, waking at night and feeling more than ordinary heat, perceived that the coverlet was on fire.

Nicolas Lémery

There are many ways that phosphorus, the 15th element in the periodic table, can kill you: it has something of a 'Jekyll and Hyde' character. On the one hand, when combined with oxygen to form phosphate, phosphorus is essential to our existence; it forms the backbone of life. On the other hand, it has been described as the Devil's element, and the darker side of its nature was well known to Agatha Christie. White phosphorus has been used in bombs, rat poison, matches and medicine, all with fatal consequences.

The 1937 novel *Dumb Witness** is the only Agatha Christie story to use phosphorus as the means of murder. On the book's very first page Miss Arundell, a wealthy old spinster, dies, apparently of natural causes. She has been in delicate health for more than a year, after a near-fatal case of jaundice. The symptoms of her final illness suggest another attack of the same liver complaint, and the cause of death is officially recorded as yellow atrophy of the liver. Nothing seems out of the ordinary until the reading of the will. The sole beneficiary is Miss Arundell's companion, Minnie Wilson, who has only recently been appointed to the role. Miss Arundell's relatives, who all seem to be heavily in debt and highly unscrupulous, have been overlooked.

Although there was nothing medically to suggest that Miss Arundell was murdered, the circumstances surrounding her death and the behaviour of her family make Hercule Poirot suspicious. Together with Captain Hastings, Poirot carries out an unofficial investigation to prove that Miss Arundell was murdered. One of the more important clues comes from a seance Miss Arundell took part in just days before her death, when a strange halo of light formed around her head. Was it an apparition of ectoplasm? A spectral premonition of death? Or the eerie glow of luminescent phosphorus from the poison she had just swallowed?

The phosphorus story
Phosphorus comes in several different forms usually distinguished by their colour – white, red, violet and black. White phosphorus was the first to be discovered, with credit usually given to Hennig Brand,[†] an alchemist working in the latter half of the seventeenth century. Others may perhaps have

---

*Entitled *Poirot Loses a Client* in the United States.
[†]*c*.1630 to either *c*.1692 or *c*.1710. The dates are hazy.

made the discovery earlier, but Brand has by far the best story. Brand was in search of gold and, for reasons best known to himself, decided to search for it in urine (perhaps the colour was the clue). After collecting many buckets of his own urine he allowed the liquid to stagnate until worms started to grow in the festering fluid. The liquid was then boiled down to a thick paste – probably by an assistant, rather than by Brand himself. The paste was heated again under a hotter flame to produce a red oil, a black spongy layer and a white solid. The white solid tasted salty, and was thrown away. Unknown to Brand, the white solid actually contained most of the phosphorus, in the form of phosphate salts. Brand continued his experiments, and combined the red oil and black layer before heating them again. From this mixture came white fumes that Brand collected in a glass vessel containing water. The fumes condensed to form a white, waxy solid. There was no gold, but what he had produced glowed in the dark. It is the glow that gives phosphorus its name, after the Greek for 'light bearer'. At this time very few substances were known to emit light without the accompanying heat of a flame, so the substance was of some worth.

Brand's method was highly inefficient, producing only a 1 per cent yield of phosphorus from the gallons of urine that had to be collected and stored. The worm stage was also unnecessary; phosphorus can be produced from fresh urine as easily, and rather less unpleasantly, as from stagnant, worm-infested urine. However, Brand had succeeded in isolating a new element, only the 13th to be discovered of the 92 naturally occurring ones in the periodic table, and the first to be found since ancient times. Brand never revealed his method for producing phosphorus, but he was happy to sell lumps of it when he was short of funds. He did, however, let on that the source of this wondrous substance was human in origin, and a few others managed to work out Brand's method for themselves.

One person who worked out how to produce phosphorus was Daniel Kraft, a renowned German chemist of the same era. Kraft started demonstrating the properties of this new substance at scientific gatherings. The lights would be turned out and small quantities of phosphorus were then smeared onto the face and the backs of the hands to demonstrate the strange glow of this substance, which was emitted without any noticeable heat. Phosphorus smeared on to a piece of paper would be warmed gently, causing the paper to catch fire where phosphorus was present. One such gathering took place for a few select members of the Royal Society in London, one September evening in 1677. Among the assembled gentlemen was Robert Boyle (1627–1691). Boyle, known today as the 'Father of Modern Chemistry', was, at that time, an alchemist, searching for the philosopher's stone.* Kraft's demonstration of phosphorus inspired and fascinated Boyle, and led him to carry out painstaking and detailed experiments to learn all he could about its mysterious properties. The systematic approach adopted by Boyle, and the accessible way he wrote up his experiments, helped him become a highly respected, world-renowned scientist. But to carry out his experiments, Boyle needed a supply of phosphorus. By 1680 Boyle had worked out the process, and collected all the urine from his house in Pall Mall, London. Together with his assistant Ambrose Godfrey (1660–1741), he processed the urine to maintain a steady supply of phosphorus for his experiments.

After leaving Boyle's employment, Godfrey continued to refine the method. Working in his laboratory just off the Strand in London, he produced the best white phosphorus, which he then supplied to Europe and beyond. Godfrey's source of phosphorus was still urine; anyone living near his lab had to put up with the smell, until after 1769, when Johann Gottlieb Gahn (1745–1818) and Carl Wilhelm Scheele (of Scheele's

---

*The mythical philosopher's stone (much popularised in recent times by a certain boy wizard) was believed to be the key to turning base metals into gold. It's what every alchemist was searching for, along with the elixir of life and the universal solvent (the last of which is particularly bewildering; if you found it, what would you keep it in?).

Green fame; see page 27) realised that phosphorus could also be obtained from bone ash.

White phosphorus is made up of units of simple molecules, consisting of four phosphorus atoms bonded together (P_4). White phosphorus is not soluble in water, but it is soluble in oils and fats. It can turn yellow when stored for a long time, so it is sometimes called yellow phosphorus, but both yellow and white phosophorus refer to the same thing. This form of phosphorus is highly reactive, particularly with oxygen, so it is stored under water. White phosphorus will easily react with oxygen in the air, burning to form clouds of white phosphorus oxides. Burning phosphorus generates an intense amount of heat as well as smoke, which led to its use in incendiary bombs and as a smoke screen in warfare.

Many people have tried to utilise phosphorus's light-emitting properties to devise safe systems of lighting. They all failed, because of the extreme flammability of phosphorus. Even if the fire risks could be reduced, the process of generating phosphorus's eerie glow also produces an unpleasant garlicky smell.* Phosphorus lighting was therefore both hazardous and unpleasant to live with. However, the flammable properties of white phosphorus were a positive benefit in its main household use, matches. White phosphorus was first used to make matches around 1830, and by the middle of the nineteenth century matches, with white phosphorus forming as much as 20 per cent of the match-head, were produced by the billion. The matches were easy to light, but also prone to igniting unexpectedly. For example, the heat of friction produced from treading on a match could cause a person to go up in flames, as happened to the 19-year-old Archduchess Matilda in 1867.*

*I worked with phosphorus-based compounds for many years; their smell is very powerful and uniquely unpleasant. The smell is often described as 'garlicky', but in reality it is like nothing else on Earth.

Indeed, carrying a box of matches in your pocket could generate enough friction from the matches rubbing against each other inside the box to cause them to catch light. Even matches left on a window sill could ignite when the sun's rays fell on them.

From the 1840s onwards, safety matches were produced using red phosphorus instead of the white form. The red phosphorus was stuck to the side of the matchbox, and the friction from running the match-head along the box would cause enough heat for the match to catch light. Red phosphorus is produced by heating white phosphorus at high temperatures (in the absence of air), causing the small P_4 molecules to link into chains. Red phosphorus is much less volatile and less flammable than white phosphorus, and non-toxic. However, it is also more expensive, so at first safety matches didn't really catch on. Concerns about the safety of white phosphorus were raised as early as the 1850s but its use in matches wasn't banned until 1906.

To produce the huge number of white phosphorus matches needed to meet demand required a large workforce, working long hours in uncomfortable and often dangerous conditions. The danger of these matches lay not just in their flammability but in the toxic nature of the white phosphorus needed to produce them. Sticks of poplar wood, twice the length of the final match, would be arranged in racks and bound tightly. It was then the task of a 'dipper' to dip both sides of the sticks in 'the compound', with the racks of matches then dried in an oven. The matches would then be cut in half and boxed up, ready to sell. 'The compound', a mixture of glue, colourant, sulfur and white phosphorus dissolved in water and heated by steam to maintain the right consistency, was held in shallow trays just the right depth for dipping the match-heads. A

*Matilda was an Austrian princess, betrothed to the future King of Italy, Umberto I. She trod on a match as she leaned out of a window to talk to a relative; her dress was on fire before she realised what had happened. Matilda subsequently died of her injuries.

skilled dipper could produce an astonishing 10 million matches in a single ten-hour shift. All the time the dippers would be breathing in phosphorus fumes, and those employed to box up the matches would be breathing in phosphorus dust.

White phosphorus is highly toxic and, in around 20 per cent of the workers, it would cause a condition known as 'phossy jaw'. This would start off as a toothache, then the teeth would fall out, and the gums, jaw and face would become painful and swollen. Slowly the soft tissue and bone were eroded away. Abscesses would appear on the gum that would ooze a most foul-smelling pus. Further abscesses would appear along the jaw line, forming a wound through which could be seen the dead bone of the jaw. The only treatment for phossy jaw was the removal of the jaw bone, replacing it with an artificial jaw. Without this procedure the phosphorus would go on to cause damage to the internal organs, leading to death. Around 5 per cent of phossy jaw sufferers died from its effects. A study in France found that half of the people suffering from phossy jaw committed suicide rather than put up with the pain and foul stench of the condition.

Other than removing an individual from the source of poisoning there is very little that can be done to treat phosphorus poisoning. Thankfully, the chances of being exposed to white phosphorus today are very low. The tragedies that occurred in match factories led to hugely improved working conditions in many industries, not just in match-making, and exposure to white phosphorus fumes in the workplace is thankfully a thing of the past.

It could take years of environmental exposure to white phosphorus before phossy jaw developed, and even then the result would be far from guaranteed so, from a murderer's point of view, something more rapid and reliable would be necessary. Phosphorus can be absorbed through the skin and gastrointestinal tract as well as through the walls of the lungs. The reactive nature of white phosphorus causes burns, so skin absorption will leave behind a nasty injury. Those working with incendiary bombs in the past had skin burns treated with copper sulfate solution. This is effective in neutralising the effects of phosphorus but copper sulfate is also toxic, and its use in this treatment has mostly been discontinued. Burns on the skin are likely to arouse suspicion, so the usual approach from a poisoner's point of view would be ingestion. If the intent is to murder an individual this is the most reliable route; even small quantities of around 100mg will be fatal.*

The toxic effects of phosphorus were recognised early on; Eilhard Mitscherlich (1794–1863), a Prussian chemist, was the first to suggest the use of phosphorus in a paste as a rat poison. He was also aware that humans might use poison intended for rats to dispose of other unwanted 'pests'. Mitscherlich wrote a paper describing a method for the detection of phosphorus in cases of possible poisoning. The method, now known as the Mitscherlich test, consists of taking the substance suspected of containing phosphorus and heating it with water in a flask. The steam vaporises the phosphorus, which is then cooled and collected in a condenser. By viewing the apparatus in a darkened room the glow of phosphorus can be observed.

How phosphorus kills
White phosphorus can burn the gullet and stomach as easily as it can burn the skin owing to its highly reactive nature. In large quantities it damages the stomach lining, causing haemorrhaging and intense pain, with vomiting of blood. The vomit may smoke as the phosphorus reacts with oxygen in the air, and there will be a garlicky smell. Phosphorus will also

---

*Eating match-heads was a popular suicide method for a while. There was enough white phosphorus in the match-heads of one box of matches to be lethal. Today, due to the use of non-toxic red phosphorus, licking the side of a matchbox, or even several matchboxes, won't prove fatal.

react with stomach acid to produce phosphine gas (PH_3), which has a particularly pungent smell that can be detected on the victim's breath. There would be a burning sensation in the mouth, and victims experience a strong thirst. Smaller quantities of phosphorus may inflame the stomach lining without causing vomiting, although a garlicky smell on the breath may still be detectable. Absorption from the gut into the bloodstream is relatively slow and can take between two and six hours, but this will be faster if the victim has eaten greasy or oily food. When the initial symptoms of pain and vomiting subside there may be a period of respite when the patient appears to get better. Unfortunately, if enough phosphorus has been absorbed into the bloodstream this is not the case; worse is to come.

From the intestines, the blood carrying the phosphorus will travel to the liver, where the phosphorus will accumulate. The liver is the main organ responsible for detoxifying the body. It is the first organ to receive blood that has come from the intestines, and is a kind of 'clearing house' for anything entering the body through the digestive system. It filters nutrients from the blood, as well as waste products or harmful substances. Some harmful substances may be excreted without change because they are already sufficiently water-soluble to be passed in urine. However, white phosphorus is not soluble in water, so enzymes in the liver carry out chemical reactions to make it more water-soluble, so that it can be excreted. The products of these reactions, like phosphorus, are very reactive and cause damage to the cells in the liver. The liver will become enlarged and jaundice will become apparent as the basic functions of the liver break down.

Jaundice is characterised by a yellow pigmentation of the skin and whites of the eyes, and it is often a sign of liver disease. It is caused by increased concentrations of bilirubin, a waste product from the breakdown of red blood cells, in the body. Red blood cells are constantly being replaced, with the old ones broken down by the liver into bilirubin, which is transported to the bile duct to be excreted in the faeces.

The kidneys may also be affected by phosphorus, as they act as a filter, removing waste from the bloodstream into urine to be excreted, but it is not kidney failure that kills in this case. When the liver is damaged so badly that it is unable to function properly, the rest of the body is subjected to high levels of poisonous compounds. By this point there is very little that can be done to treat the patient, beyond supportive care. Death from liver failure ensues three or four days after phosphorus has been ingested.

Some real-life cases

There have been many real-life cases of murder and suicide by phosphorus, but a couple of cases have some interesting parallels with the plot of *Dumb Witness*. The book was written in 1937, and both of the following real-life cases occurred in the 1950s.

In 1953, Louisa Merrifield and her third husband, Alfred, moved in with 80-year-old Sarah Ann Ricketts as live-in housekeepers. Mrs Ricketts was healthy for her age but had trouble walking, and she seldom went more than a few steps from her front gate. She needed help with day-to-day activities such as cleaning, shopping and preparing meals. From the start it is clear that employer and employees did not get along, and Mrs Ricketts complained to friends and neighbours that she was not being given enough food.

At 6.30 p.m. on 13 April 1953, a doctor attended Mrs Ricketts and found nothing wrong with her, but the following day Dr Yule, her usual doctor, was called urgently. When he arrived shortly after 2 p.m. Mrs Ricketts was dead. Dr Yule could not account for her death, as he had not attended the patient recently in a professional capacity. However, he had spoken to her for fifteen minutes only four days previously, and she had seemed in her normal good health. Dr Yule declined to write a death certificate, and reported the death to the Coroner's office.

A post-mortem was ordered that revealed Mrs Ricketts's liver was 'the colour of putty, with a cheesy consistency'[*]

---

[*] A quote from the pathologist.

consistent with phosphorus poisoning. The appearance of the liver confirmed that the poison had been administered recently, rather than in small quantities over a period of time. The pathologist could not confirm where the phosphorus had come from, but he suggested the most likely source was a tin of Rodine rat paste. The other ingredient in Rodine rat poison was bran that was mixed with the phosphorus; since bran was found in Mrs Ricketts's body and in close association with phosphorus, the pathologist believed the two must have been consumed at roughly the same time. No Rodine rat poison was found at Mrs Ricketts's bungalow or among the Merrifields' possessions. Rodine could be easily purchased from chemists, but even though it contained a lethal amount of phosphorus, there was no legal obligation to sign the poison register. The police therefore were unable to trace the purchase or any poison to the Merrifields.

However, an incriminating piece of evidence was found in Louisa Merrifield's handbag – a dirty spoon that had some congealed material stuck to it. It seemed an unlikely item to carry around in a handbag, but analysis of the congealed material revealed no trace of phosphorus. However, in the time between its possible use to mix the poison with food and its analysis in a forensic laboratory the phosphorus could have evaporated. It was suggested that the congealed material could have been jam, or it could have been rum and sugar, which Mrs Ricketts was known to take. Mrs Ricketts was also given brandy during her final illness by the Merrifields. Tests with brandy showed it was quite effective at masking the flavour of phosphorus, but it could not completely mask the smell; blackcurrant jam was effective at masking both the taste and the smell.

With clear signs of phosphorus poisoning at the post-mortem but no evidence of how it was administered or from where it had been obtained, it appeared that the case against the Merrifields was pretty thin. However, two weeks before her death and less than three weeks after the Merrifields had moved in, they had somehow persuaded Mrs Ricketts to change her will in their favour. The day before she died a tradesman working in the bungalow heard Mrs Ricketts ask Alfred Merrifield to call at her solicitors as she wanted to change her will. She remarked to the tradesman, 'They are no good to me. They will have to get out.' Louisa Merrifield also slipped up when she told an acquaintance that she had been living with an old lady who had died and left her a bungalow worth £4,000, even though Mrs Ricketts was, at that point, very much alive and in good health.

The jury deliberated for five and three-quarter hours before finding Louisa Merrifield guilty and Alfred Merrifield not guilty. Louisa was hanged, but Alfred was free to inherit a half-share of the bungalow left to him in Mrs Ricketts's will.

A second murder case shows how easily phosphorus poisoning can be mistaken for natural causes. In 1954, in a desperate attempt to avoid being forced into a second unwanted marriage by her parents, a young widow walked into a police station and confessed to murdering her first husband. She claimed that three and a half years previously she had given her husband phosphorus from a tin of rat poison. Initially the woman wasn't believed. Her first husband was thought to have died of natural causes, and she was not thought intelligent enough to have carried out the plan without being detected. Despite some misgivings the matter was investigated further, and it was found that the woman's husband, aged 35 when he died, had, until the day of his alleged poisoning, been in excellent health. The symptoms of his illness, attributed to haematemesis (vomiting of blood) from a gastric ulcer, included vomiting and severe abdominal pain before a brief recovery. Two days later the symptoms returned but worse than before, and they were accompanied by a great thirst. He died later the same day after vomiting large amounts of blood.

The body of the husband was reluctantly exhumed. It was felt that after three and a half years there would be little evidence of poisoning remaining, but the body was in good condition and showed no natural cause for the symptoms displayed by the victim in his last few days. Tests on a number of organs and tissues, using the procedures outlined by Mitscherlich, confirmed the presence of elemental phosphorus. The court decided that murder had been proven, but the widow was not held responsible for her actions.

Agatha and phosphorus

In *Dumb Witness*, the death of Miss Arundell does not raise any suspicions with her doctor, family or friends, and is attributed to yellow atrophy of the liver, a condition she had been suffering from for several years. No one except Poirot is suspicious of the death; his real challenge is to prove that Miss Arundell was murdered. Agatha Christie highlighted the difficulty, which remains a problem today, of proving guilt in cases where the murderer has used poison. First, it has to be established that the victim did not die of natural causes.

The murderer in *Dumb Witness* chose their poison carefully. Jaundice is a symptom of phosphorus poisoning, and so mimicked a recurrence of Miss Arundell's liver complaint. Although the exact cause of Miss Arundell's complaint was never revealed, we know she had a near-fatal attack of jaundice eighteen months before she was poisoned. Miss Arundell appeared to have recovered from her earlier illness but was taking medication to manage her condition. She was also careful about what she ate and avoided rich foods, though she occasionally lapsed into bad habits. On the evening that Miss Arundell first developed symptoms of her final illness she had eaten a curry. The curry may have triggered an attack of liver trouble, but it would also have been the perfect vehicle for administering a poison such as phosphorus. The flavour of the curry would be strong enough to disguise the distinctive taste of phosphorus; curry can also be greasy, which would increase the rate of absorption of phosphorus into the bloodstream.

Anyone who eats a lot of phosphorus may excrete it unchanged in their faeces because it has stayed in the gut and therefore hasn't been processed by the liver. This can result in 'smoking stool syndrome', where white smoke emanates from a patient's waste as the phosphorus reacts with oxygen in the air. The faeces may even glow in the dark. Christie is too discreet to discuss something as unpleasant as the condition of Miss Arundell's bowel movements, but the fact that the doctor attending her never considered phosphorus poisoning indicates that Miss Arundell did not suffer from this condition.

A post-mortem examination would have been expected to reveal the true cause of death, but none was carried out because Miss Arundell was thought to have died of a known illness. Poirot considers having the body exhumed, but this would have been risky. It would have upset the family, and it would have been against the wishes of the deceased; plus it might not have led to any positive results after the body had been buried for two months. Phosphorus might leave clear signs of damage to the internal organs, such as haemorrhage or lesions from the corrosion caused by the phosphorus reacting with tissues in the body, but it might not. A large dose of phosphorus may be detectable initially by its characteristic smell when the stomach is opened; a pathologist may even be able to see the characteristic eerie green glow from the intestines if the lights are turned out. The longer the body has been buried, though, the more likely it is that the phosphorus has been converted into phosphate by reaction with oxygen. The glow would no longer be present, and phosphorus oxides from the poison would be indistinguishable from similar compounds naturally found in the body.

Damage to the liver is usually a better indicator of phosphorus poisoning, though probably not in Miss Arundell's case. Together with other indicators in the body, the liver can give clues as to when the poison was administered and whether phosphorus was ingested in a single large dose or in smaller doses over a long period of time. In the case of a very large dose of phosphorus, death may occur as rapidly as 12 hours later, due to the direct action of phosphorus on heart muscle resulting in cardiovascular collapse. If death had occurred several days after ingestion, as in the case of Miss Arundell, the liver would be enlarged with signs of jaundice – the kind of damage you would see from a naturally occurring liver disease. The longer the victim survives after being poisoned, the more likely it becomes that the damage will spread from the liver to the kidneys, where phosphorus metabolites produced by the liver are filtered from the blood into the bladder for excretion in the urine. If the poisoning had been chronic and Miss Arundell had received regular small doses of phosphorus over a long period of time, her liver may have looked more like that of an alcoholic (i.e. enlarged with signs of jaundice), and therefore indistinguishable from the effects of her medical condition.

Even without the possibly dubious results of a post-mortem, Poirot finds other evidence that phosphorus has been given to Miss Arundell. A few days before her death, and the night Miss Arundell started to feel unwell, she took part in a seance. During the seance a ribbon of glowing vapour was seen to issue from Miss Arundell's mouth, forming a halo around her head. The halo was witnessed by several people but attributed to ectoplasm, or some spectral manifestation. Only Poirot realises it was the green glow of phosphorus vapour.

Poirot describes how the halo of light was some 'phosphorescent' substance; he misuses this term, which technically refers to the glow emitted by a substance that has been charged by light shining on it, and continues to glow in the dark (exit signs in cinemas, for example), although the word does have its origins in the glow observed when phosphorus is exposed to the air. Phosphorus actually *chemiluminesces*; light is produced as the result of a chemical reaction (other examples of this include the bottoms of female fireflies, or the glow-sticks waved by night-clubbing ravers). The exact cause of the glow emanating from white phosphorus was not fully explained until 1974, so Poirot can hardly be blamed for his mistake back in 1937. A chemical reaction between phosphorus atoms and oxygen atoms produces diphosphorus dioxide (P_2O_2) and monophosphine oxide (HPO) molecules that only exist fleetingly, but both of which emit light. A piece of white phosphorus will glow in a stoppered flask until either all of the oxygen or all of the phosphorus has been consumed in the reaction. Looking back at the spectral halo, body temperature would have increased the amount of vapour produced from the phosphorus in the stomach, and as Miss Arundell breathed out, the vapour may have been visible as it reacted with the oxygen in the air and began to glow.

Confusing the glow of phosphorus for ectoplasm at a seance is not as far-fetched an idea as it may sound. Many people have tried to explain unusual phenomena such as graveyard ghosts, will-o'-the-wisps and spontaneous human combustion as phenomena caused by phosphorus, either from the glow of elemental phosphorus reacting with oxygen, or through a phosphorus-based compound spontaneously catching fire in the air. It has long been known that in marshes, bogs and other areas where organic matter decays, methane (CH_4) is produced by bacteria in the anaerobic (oxygen-free) conditions found in the damp soil. Methane burns very effectively but it needs a source of ignition, which could be from phosphine (PH_3) or diphosphane (P_2H_4); these compounds can also be produced by anaerobic bacteria, and they spontaneously combust when exposed to air. The theory goes that phosphine or diphosphane is slowly emitted from the soil, catches fire and triggers the burning of methane to produce the elusive flickering flames known as will-o'-the-wisp, or *ignis fatuus* (foolish fire). Similarly, the reaction of phosphorus with stomach acid could produce phosphine gas, which could react spontaneously with the air when breathed out.

Even if the real reason for Miss Arundell's final illness had been determined before she died, very little could have been done. In the early stages of phosphorus poisoning it is hoped that the vomiting will remove most of the poison from the system, and water can be given to the victim between bouts. Once Poirot is convinced that Miss Arundell died of phosphorus poisoning, he sets about proving how it was done. Miss Arundell's symptoms began in the late evening, indicating that she had been poisoned around dinner time. Poirot questions the nurse who attended Miss Arundell in her last illness to find out who prepared her food, and who had access to the sick room. It was found that the prime suspect, a faithful companion, never went near the food and was rarely in the sick room. Only the servants and the nurse would have been in a position to administer poison in the food, and they were all above suspicion.

Miss Arundell was in the habit of taking two or three pills after each meal. Could these have been the source of poison? We were not told what medicine Miss Arundell's doctor prescribed to her, but we are assured it was mild. We also learn that Miss Arundell was taking a patent medicine, Dr Loughbarrow's Liver Pills, which contained aloes (used as a herbal medicine, usually a laxative) and podophyllin (once used as a resin to treat genital warts). Despite the doctor's assurances, podophyllin is quite a toxic substance, and overdoses can lead to depression and even death. It is used mostly in patent medicines to treat constipation or to increase secretion of bile, but it is not recommended for anyone suffering from jaundice. Christie seems unaware of the complications of podophyllin, and considers these pills to be innocuous.

Phosphorus has been used as a medical treatment in the past; its popularity waxed and waned over a period of 300 years before it was finally abandoned. The source of phosphorus and its glow were seen as evidence of the vital 'flame of life', as Robert Boyle put it – the *flammula vitae*. Phosphorus pills, carefully coated to prevent them from reacting with the air, were prescribed in the early eighteenth century for colic, asthmatic fevers, tetanus, apoplexy and gout. None of these treatments would have worked, and the larger doses of phosphorus would kill. The dangers were soon realised, a mere 90 or so years later, and phosphorus started to disappear from the regular pharmacopoeia. However, it was still used, in ever-lower doses, for the treatment of conditions as varied as tuberculosis and epilepsy. In the early twentieth century phosphorus was still being prescribed as a general tonic, but by 1932 it had disappeared from the *British Pharmacopoeia,* though it persisted in some remedies until the 1950s. When *Dumb Witness* was written in 1937, phosphorus would not have been prescribed and, even if it was found in over-the-counter remedies, the dose would be in the region of 0.5–3mg, well below the lethal dose. These pills would gradually become safer the longer they were left on the shelf or in the medicine cabinet, as more and more of the white phosphorus was oxidised into safer compounds over time.

Dr Loughbarrow's Liver Pills are an invention of Agatha Christie, but very similar pills were sold for neuralgia or as a general tonic. In 1931 a Dr G. Coltart wrote in the *Lancet* about a case he had prescribed for in 1904. A patient had come to him complaining he felt 'run down', so a prescription for a common tonic pill, which contained strychnine and phosphorus, was made out for him. The patient was told to stop taking the pills if the strychnine made him twitch. Twenty-seven years later, the patient returned with a severe case of phossy jaw. Asked why he had continued to take the pills he said they had never made him twitch.

There was no phosphorus in Miss Arundell's medicine, but the medicine could have been tampered with. The liver pills were in the form of gelatin capsules, two halves fitted together to enclose the powdered medicine inside. All the murderer had to do was get one of the capsules, empty it of its normal herbal remedy and add powdered white phosphorus in its place. A lethal dose of 100mg of phosphorus would easily fit inside such a capsule, and more could be added just to make sure. A single adulterated pill added to a box of 50 pills could be taken at any time, and the murderer was unlikely to be present.

As Poirot points out, phosphorus was easy to obtain from rat paste. Rodine rat poison was sold in pharmacies across the UK, and a one-ounce tin contained ten grains of phosphorus (approximately 650mg), enough to kill six people. Poirot also suggests that a trip abroad might provide an even easier source of phosphorus, as white phosphorus-based matches continued to be sold in many places long after they had been banned in Britain. Poirot, of course, pieces together all the circumstantial evidence, exposing the murderer and allowing justice to take its course.

RICIN

Partners in Crime

Don't Torture Your Child!
TO MOTHERS! See your little one's terror at the very thought
of a dose of castor oil, mineral oil, calomel or pills. Urgh!
Advert for 'Cascarets Candy Cathartic', 1918

Many people will have first-hand knowledge of the unpleasant nature of castor oil, but fortunately very few will experience the deadly effects of another extract from the same plant – ricin. This poisonous protein has considerable notoriety today, but when Agatha Christie was writing it had never been used in a murder. The lack of real-life examples may underlie some of the inaccuracies in Christie's use of the poison, a rare occurrence for the Queen of Crime. Before 1978 ricin was untreatable and untraceable, seemingly the perfect poison. In many respects Christie was years ahead of her time when she used it to poison four members of the same household in *The House of Lurking Death*, a short story featuring two detectives, Tommy and Tuppence Beresford, which along with a number of other tales featuring the duo appears in Christie's 1929 book *Partners in Crime*.

The House of Lurking Death begins with Lois Hargreaves asking Tommy and Tuppence to look into an attempt on her life made via some poisoned chocolates. The arsenic-laced chocolates had been sent anonymously, but Lois could prove that they must have been sent by someone in her household. Tommy and Tuppence promise to travel to Miss Hargreaves's house, Thurnley Grange, the next day to investigate. But before the pair can set off, another poisoning attempt is made. A plate of ricin-laced fig sandwiches is served for afternoon tea; this time the murderer is successful.

The ricin story

Ricin is produced by the castor oil plant, *Ricinus communis*, which occurs throughout tropical regions of the world and is often grown as an ornamental shrub. Small amounts of ricin can be found in all parts of the plant, but it is mainly concentrated in the seeds; specifically, it occurs in the endosperm, a store of oil that serves as a source of nutrition for the fertilised seed. Ricin is important for the plant, as it deters herbivorous animals from eating it (although birds such as hens and ducks appear immune to the castor oil seed's poisons). Between five and twenty raw castor oil seeds may prove fatal for an adult human if they are well chewed, but cooking the seeds inactivates the poison; anything that changes the shape of a protein is said to 'denature' it. This can be done by heat or chemical reaction, and it is irreversible – think of the result when cooking eggs, which are primarily made up of proteins.

Most of us will recognise castor oil as a safe and relatively mild laxative, sold as an over-the-counter remedy, but castor oil and its derivatives have a wide range of uses as lubricants, in paints and dyes, in plastics, pharmaceuticals and many more; it is ricinoleic acid, a fatty acid found within castor oil, that is of particular interest industrially because of its chemical properties.

The plants are grown widely as a crop to satisfy the considerable demand for castor oil and ricinoleic acid. The oil is pressed from the seeds, and the remaining husks, or mash, can contain up to 5 per cent ricin. The mash can be used as a fertiliser but not as cattle feed, as is often the case with husks of other seeds from which oil is extracted, because it would poison the animals. Ricin is water-soluble but will not dissolve in fats, so little is found in the oil after it is pressed from the seed, but to ensure no ricin makes it into the castor oil it is heated to more than 80°C as it is extracted; this denatures the protein, so inactivating it.

Harvesting castor oil seeds can be hazardous to those who come into contact with the seeds, though not necessarily because of the presence of ricin – the tough outer shell of the castor seeds prevents the ricin from being released. Swallowing the seed whole without chewing is unlikely to prove fatal, as the seed coat is tough enough to survive the digestive process intact. However, the plants have allergenic compounds on their surfaces, which can cause permanent nerve damage to those who work with the plants. For reasons such as these there is great interest in finding alternative sources of ricinoleic acid, or in genetically modifying castor oil plants so that they do not produce ricin or these allergenic compounds.

How ricin kills
Ricin is a toxalbumin, a poisonous protein formed from two chains, A and B, which are linked by a single bond between two atoms of sulfur. All cells have a cell membrane that controls which substances are allowed to pass in and out of the cell. Chain B of ricin attaches to the surface of a cell, allowing the whole ricin protein to cross the cell membrane and enter the cell body. Inside the cell the two chains break apart,* releasing chain A, which is the part that does the damage.

*This is due to an enzyme in the cell, protein disulfide isomerase. The body has to split up disulfide binds regularly in the course of its normal function, so it is not surprising that there's an enzyme to do this.

Ricin is classed as a ribosome-inhibiting protein, which is abbreviated to RIP.* When separated from chain B, chain A can permanently disable the ribosomes, structures within cells that build new proteins and enable cells to function and replicate. Chain A breaks a crucial bond in the ribosome structure, inactivating it. Since ribosomes are essential for metabolism, growth and repair in the body, this is very bad news; the body can no longer carry out basic processes, or repair itself. If enough ribosomes are inactivated a cell will die, and if enough cells die in an organ, haemorrhage occurs and the organ fails. In addition, a single molecule of chain A is not restricted to damaging just one ribosome – it can roam around the cell, disabling up to 1,500 individual ribosomes per minute without being damaged or destroyed itself.†

The ability of one molecule to do so much damage means that ricin is lethal in extremely small doses. Less than 1mg injected under the skin is enough to kill an adult human; inhaling ricin dust can be just as deadly. Ingestion of ricin is slightly less dangerous, because the gastrointestinal tract treats ricin like any other protein we eat and breaks it up into its constituent amino acids for use in the body, thereby inactivating it. A murderer would need at least 100 times more ricin in order to kill by ingestion; however, this does still equate to just 100mg to kill an adult.

The symptoms of ricin poisoning appear approximately six hours after ingestion, and only slightly quicker if the poison is administered by injection or inhalation. There is a sensation of burning in the mouth, followed by nausea, vomiting, cramps, drowsiness, cyanosis (blue colour in the skin), stupor, circulatory

---

*Appropriately.

†The two protein chains that constitute ricin occur in many other plants, but not together. Plants that contain only chain A, barley for example, are safe to eat even though they contain the poisonous half of the ricin protein, because without chain B, chain A cannot get inside the cells to do any damage.

collapse, blood in the urine, convulsions, coma and death. Haemolysis (the breakdown of red blood cells) occurs even at extreme dilutions of ricin, resulting in severe haemorrhaging. Death occurs between three and five days after poisoning.

Ricin has no approved therapeutic uses, but there have been suggestions that it could be modified into an effective treatment for cancer, because of its ability to disrupt cell function. Another potential use would be to use chain B to deliver pharmaceutical drugs into a cell interior (since without chain A, chain B would have no damaging effects).

Despite the fact that castor oil plants can be grown legally, researchers working with ricin must be registered if they are storing more than 100mg in their lab. Outside of registered laboratories, anyone trying to isolate ricin or with pure ricin in their possession would certainly require a very convincing argument for the authorities. The high toxicity of ricin and its relative availability means it has been looked at closely as a potential biological weapon. Trials were conducted by the United States Army during the First World War; it was concluded that ricin was no more effective than phosgene or the other chemical-warfare agents that were then in use, the major problem being that methods of deploying ricin created sufficient heat to denature the protein. In the years leading up to the Second World War and during the conflict itself, more research was conducted independently by the Allies. The French thought ricin too dangerous for research until an antitoxin had been developed, so their plans were abandoned at an early stage. The British got further in their research and developed bomblets that were designed to spread clouds of inhalable ricin dust. The US Army's research was even more advanced; it developed the means to produce large quantities of ricin powder, or 'Agent W', using a novel chilled-air grinder to prevent the heat from the friction of grinding from denaturing the protein.

Field trials of 'Agent W' exposed another downside of ricin as a biological weapon. Any area exposed to ricin dust remained dangerous for a long time (until the bond between chain A and chain B had been broken – this could be two or three days). The dust would also attach to clothing and could subsequently be inhaled, proving fatal to friend and foe alike. The delayed action of the poison and ethical concerns over its use meant that ricin was never used in combat. The protein is less stable than anthrax and less toxic than botulinum toxin (which was also developed, as 'Agent X'), so a much higher quantity must be produced to guarantee the same killer effect.

Is there an antidote?

There have been attempts to develop antidotes for ricin poisoning because of the poison's potential use as a biological weapon, but nothing is commercially available to date. Supportive care is the best that can be offered to anyone unfortunate enough to be exposed to ricin.

However, there are preventative measures that can be taken before exposure, in the form of a vaccine. This is developed from an inactive form of chain A, and it has been shown to be effective for several months – though you need a little bit of time for the body to develop its own antibodies. Because ricin is quickly cleared from the blood (owing to its entering the cells), treatment with antibodies after exposure is ineffective – this would be like giving someone a measles vaccine after they have already come out in spots.

Some real-life cases

Most incidents of ricin poisoning are due to the accidental ingestion of castor oil seeds;* the survival rate is an excellent 95 per cent because the seeds can generally be removed from the

---

*The pretty mottled beans are sometimes picked and eaten by children, who obviously don't know about their deadly properties. Apparently they taste like hazelnuts; I don't recommend finding out for yourself.

stomach before much of the ricin has been absorbed. The first ricin murder in Britain took place in 1978, after Agatha Christie's death, but it is one of the most famous cases of assassination.

Georgi Markov was a Bulgarian dissident, writer and journalist working for the BBC and Radio Free Europe. He made many broadcasts about the regime in Bulgaria and the privileged lives of those in power in that country, and was particularly scathing in his comments about the president, Todor Zhivkov. On 7 September 1978, as Markov waited at a bus stop on Waterloo Bridge, he felt a sharp pain in his thigh, like an insect sting. Turning around, he saw a man bending over to pick up his umbrella; the man apologised and quickly crossed the road to get into a taxi, which drove away. Markov later told one of his colleagues at work about the incident, and showed him the tiny mark on the back of his thigh. Markov recorded his broadcast, and went home a couple of hours later.

That night Markov developed a fever and started vomiting. The next day he was still feeling very unwell, and he stayed at home. His doctor came to see him and called for an ambulance. At the hospital, Markov showed the house physician the wound on the back of his leg. The doctor examined the area, which was now inflamed, with a tiny pin-prick at the centre. He did not think this was the cause of Markov's illness, but the thigh was X-rayed just in case. Nothing showed up.

Markov's condition continued to deteriorate. Doctors suspected his symptoms were caused by severe septicaemia. He had a rapid pulse, very low blood pressure and an elevated white blood cell count, indicating an infection of some kind; he was not passing urine, because his kidneys were damaged, and he was still vomiting blood. Fluid started to collect on his lungs, and on 11 September his heart stopped. Three days after the incident on the bridge, Markov was dead.

Because of Georgi Markov's unusual symptoms and his sudden death, a post-mortem was carried out. This showed damage to the lungs, liver, intestines, lymph glands, pancreas and testicles, and there were haemorrhages to several organs.

Markov had died of acute blood poisoning, but the cause of the poisoning was not known. Examination of the wound on Markov's leg, however, revealed a tiny metal pellet inside the flesh (checking the X-ray again, what was thought to have been a blemish on the film was now revealed to be the pellet). Measuring only 1.7mm in diameter, the pellet had minute holes drilled into it. It was suggested that the minuscule reservoir formed where the holes met had contained the poison that had killed Markov, but there was no poison remaining in the pellet. The cavity in the pellet was much too small for a lethal dose of arsenic or cyanide; something far more potent had been used. The cumulative evidence from the post-mortem and the symptoms Markov displayed during his final illness suggested ricin, and this was the verdict given at the inquest. Further testing was carried out at Porton Down, a government-run military scientific research facility near Salisbury, England, to confirm this.

By measuring the space inside the pellet, Porton Down scientists were able to calculate the volume of poison it would have contained. They found that a mere 0.5mg of material would have filled the cavity. Ricin is one of the few poisons potent enough to kill in the quantities that could fit inside the pellet.

An identical amount of ricin was given to a pig, and the symptoms the animal developed matched the symptoms displayed by Markov. Pigs are often used as proxies for humans as they are of a similar size, they are hairless, their guts contain similar bacteria and their organs are of a comparable size, so injury and decay processes are also comparable. A post-mortem on the pig revealed similar damage to the internal organs; in fact the similarities were close enough to exclude all other possibilities. Ricin was confirmed as the poison that was used to kill Georgi Markov. The capsule could have been coated in wax, which would have melted at body temperature to release the poison, or in sugars that would have dissolved inside the body.

A similar assassination attempt in Paris, involving another Bulgarian defector, Vladimir Kostov, allowed the British authorities to make further comparisons. Ten days before the attack on Markov, Kostov had been standing on an escalator of the Paris Metro when a pellet was fired into the small of his back. He felt a sting just above the belt of his trousers. A small wound on his back appeared and swelled slightly. The following day Kostov felt unwell; he developed a fever, and went to his doctor. The doctor said there was nothing to worry about, and after a few days his temperature was almost back to normal. After Markov's death Kostov allowed the wound on his back to be examined, and a pellet, identical to Markov's, was removed, along with a sample of tissue for examination. The pellet still had some of the original coating attached, and most of the ricin was still within the pellet. The small amount that had leaked into the surrounding tissue had caused Kostov's symptoms, but it also allowed his body to develop antibodies to the toxin. These antibodies were detected in the tissue samples.

From all the evidence it was hypothesised that an agent of the Bulgarian secret service had fired a pellet into Markov's leg that contained a lethal dose of ricin. The pellet was fired either from an ordinary air pistol, and an umbrella was dropped at the same time as a distraction, or from a specially adapted umbrella designed and built with the collaboration of the Russian secret service. After the fall of the communist regime in Bulgaria it was hoped that the mysteries surrounding Markov's assassination would be cleared up. Documents relating to Markov and the incident were kept in the Bulgarian secret service archives, but many are missing or destroyed. There are still many questions to be answered; no one has ever stood trial for Georgi Markov's murder, or even been seriously accused of committing the crime.

Agatha and ricin
In *The House of Lurking Death*, Tommy and Tuppence's investigation into an attempted arsenic poisoning case quickly

escalates into finding a mass-murderer. The morning after Lois
Hargreaves had engaged the detective duo, Tommy reads in the
paper of a fatal case of poisoning at her home, Thurnley
Grange. Miss Hargreaves died less than 24 hours after voicing
her concerns about being poisoned. There was a second fatality,
a parlour maid called Esther Quant, and two other members of
the household, Dennis Radclyffe, Miss Hargreaves's cousin,
and Miss Logan, a distant relative of Dennis, were seriously ill.
Tommy and Tuppence hurry down to Thurnley Grange to
find out what is going on.

By the time Tommy and Tuppence arrive at the house
Dennis Radclyffe has also succumbed to the effects of the
poison, but Miss Logan is still clinging to life. The source of the
illness seems to be a plate of fig-paste sandwiches served at tea
the previous afternoon. The cause is initially thought to be
ptomaine (or food) poisoning, of a particularly virulent kind.
The typical symptoms of food poisoning, vomiting, diarrhoea
and stomach pain, would also be displayed by victims of ricin
poisoning. Because of the earlier attempt on Miss Hargreaves's
life, the doctor who treats the victims, Dr Burton, is suspicious
of foul play, and the fig paste is sent for analysis. While they
wait for the results to come back, Tommy speculates with Dr
Burton about arsenic being added to the fig paste, because this
poison had been added to chocolates in the earlier murder
attempt. Dr Burton dismisses this theory as arsenic would not
have killed so quickly. His initial theory is that a powerful
vegetable-based toxin has been used.

Whatever the poison was, it killed its victims within 12
hours. Tea would have been served at around four o' clock in
the afternoon, and both Lois Hargreaves and Esther Quant
must have died during the night before the newspaper deadline,
otherwise Tommy and Tuppence couldn't have read about it in
the morning paper. Twelve hours is unusually quick for ricin,
which normally takes three to five agonising days before the
victim finally succumbs. Perhaps the murderer added a
particularly high dose to the fig-paste sandwiches? Ricin,
which does not have a particularly strong flavour even in a

very large dose, would probably be masked by the flavour of the figs.

Later, the doctor sends a note to Tommy and Tuppence, saying that he has 'reason to believe that the poison employed was ricin'. There are no details telling us what these reasons might have been, and no description of any test having been carried out, but then in 1929 there *was* no test for ricin. The victims at Thurnley Grange had inflamed gastrointestinal tracts, and exhibited haemorrhaging; it was probably these symptoms that indicated ricin. There had been no cases of murder by ricin poisoning in Britain at the time Agatha Christie was writing this book, but there would have been cases of accidental ingestion of the seeds, and the symptoms and signs at post-mortem would have been known from these cases. Without a specific test being available, however, the best that could be done to confirm ricin as the poison would have been to inject an animal with ricin, or to add it to the animal's feed, and then compare the symptoms and post-mortem characteristics. This would take considerably longer than Christie allows for in *The House of Lurking Death*.

As if to confirm the suspicions of ricin, Tuppence remembers seeing some castor oil plants in the garden at Thurnley Grange. A book is also found in the house, *Materia Medica*, a work that collects together information about substances with therapeutic properties. The book is found open on the page for ricin, and contains enough information to provide a method for extracting ricin from the seeds of the castor oil plant. This seems an unusual thing to include in a book intended for medical benefit, as ricin has never been used as a therapeutic agent. Castor oil has been used for thousands of years as an emetic, a laxative, and even as an anti-dandruff treatment, and details of how to extract castor oil would presumably include methods for ensuring ricin was *not* extracted along with the oil.

The third victim at Thurnley Grange, Dennis Radclyffe, presents Tommy and Tuppence with a problem as they try to crack the case, as he was not at home when the fig-paste sandwiches were served. It is clear that Dennis had been poisoned with the same poison as the others; he was also taken ill the same night as the others, and by five o'clock in the morning he had also succumbed to the effects of poison. Dennis was spotted drinking a cocktail before dinner, and it was shortly after this that he complained of feeling ill. The glass was found and sent for analysis, which confirmed that ricin was present (although again, how this was achieved is not clear). Also, the time between ingestion and death is again unlikely, as symptoms of ricin poisoning would not have been expected to present themselves for about six hours after ingestion, with death occurring several days later.

Adding ricin to a cocktail is not the best way to poison someone; the high alcohol content in the drink would probably denature much of the ricin protein. Alcohol causes proteins to unfold. Their three-dimensional shape is changed considerably, so the protein is no longer able to carry out its function.* A cocktail such as a traditional martini would have 30 to 40 per cent alcohol, which would be expected to denature a large proportion of any protein such as ricin. There would have to be a very large dose of ricin in Dennis's cocktail glass to ensure that enough protein survived the alcohol and digestive processes to be absorbed into the body and cause such a rapid death.

The next five paragraphs contain significant spoilers. If you do not want to know who did it, look away until page 237 (and if you wish to read *Partners in Crime* – and I hope you do – this crime is just one of the 14 mysteries that make up the book).

The fourth victim at Thurnley Grange, Miss Logan, who also ate the fig-paste sandwiches, was still ill but was expected

---

*This is why alcohol gels are used to sanitise hands. At concentrations of 70 per cent, alcohol can be absorbed through the cell membranes of bacteria, denaturing the proteins inside and killing the organism.

to make a good recovery. Victims of ricin poisoning who survive beyond five days are usually expected to recover, so it may have been a little premature to give Miss Logan such a positive prognosis after only 24 hours. More importantly, how did she manage to survive for so long when the other members of her family died within hours? Maybe she didn't eat as many sandwiches as the others. Or perhaps there was another reason, as Tuppence would discover.

When interviewing Miss Logan, Tuppence notices some marks on her arm that look like pin-pricks, or the marks of a hypodermic syringe. Thoughts of Miss Logan being a morphine or cocaine addict are quickly dismissed because 'her eyes were all right'. So what was the cause of the marks? It turns out that Miss Logan had been deliberately injecting herself with small amounts of ricin.

Miss Logan's father was a pioneer of serum therapeutics, a vaccination that protects an individual from a particular disease or toxic substance (known collectively as antigens). Agatha Christie correctly states that ricin was used as a tool in the early days of research into immunology. Injections of small amounts of some toxic substances can be used to build up a natural immunity to larger, normally lethal doses. By introducing the immune system to non-lethal levels of a toxic protein, antibodies are developed and a 'memory' formed of the toxin. The antibodies are very specific to the molecule they bind to, and a body will build up a library of antibodies in response to exposure to a wide range of antigens over the course of its lifetime. An antibody binds tightly to its antigen, preventing it from entering cells in the body; it then stimulates its removal by cells called macrophages, which engulf and destroy foreign bodies, and they also trigger further immune responses. Hypodermic injections of small amounts of ricin would enable the body's immune system to produce anti-ricin, so were a larger, normally lethal, dose to be given, the body could rapidly inactivate the ricin, and the person would survive.

The body's response to the presence of ricin can now be used for its detection in poisoning cases (a test that post-dates

Christie's writing of the book, of course). Ricin antibodies can be isolated and modified for use in laboratory tests called immunoassays. Anti-ricin can be radiolabelled (tagged with radioactive particles), or modified to emit light when ricin binds to it, indicating that ricin is present in a sample. Many such antibodies have been developed, binding to various drug molecules and allowing the rapid screening of samples for multiple compounds. Other techniques, such as chromatography, can be used to identify other compounds that occur in castor oil plants, and these can be used as proxy markers for ricin exposure. The incredibly small quantities involved in ricin poisoning continue to challenge the limits of detection, though, and there is still no standard, approved method for detecting ricin, either in the environment or in human tissue.

In *The House of Lurking Death*, Miss Logan must have started her preparations for murdering her relatives weeks or even months before adding poison to the fig-paste sandwiches. Extraction of ricin from the castor oil plants in the garden could have been carried out at almost any time, as ricin can easily be stored and stockpiled for the right occasion. Miss Logan then injected herself regularly with tiny amounts of ricin, to give herself immunity. One day she added ricin to the fig-paste sandwiches served at tea, and was able to eat some, safe in the knowledge that, although she might be ill and therefore unlikely to be suspected, she would not die. Later in the evening, she followed up by adding ricin to Dennis's cocktail glass. Esther the parlour maid was not an intentional victim. She simply made the mistake of eating a sandwich on the sly – with death her ultimate punishment.

The Mysterious Affair at Styles

A final convulsion lifted her from the bed, until she appeared to rest upon her head and her heels, with her body arched in an extraordinary manner.

Agatha Christie, *The Mysterious Affair at Styles*

The Mysterious Affair at Styles was Agatha Christie's first novel, and it contains all the ingredients for her classic style of detective fiction. We have a lethal poison, a brilliant detective, a bumbling assistant, a police inspector getting everything wrong, and all in a lovely country house setting; there is even a false beard to try to throw us off the scent. This is the book that introduced the world to Hercule Poirot, and saw the dramatic demise of Mrs Emily Inglethorp in a sinister case of strychnine poisoning. With only one victim you might think this a straightforward case but there are plenty of suspects, from the gold-digging younger husband to inheritance-hungry offspring and vengeful staff – we are spoilt for choice of motive and opportunity. Among the suspects is a world expert in toxicology, a nurse, a doctor and a young woman working in a hospital dispensary. Only one thing seems certain: that the victim was dispatched with strychnine – but the real mystery is how, and by whom. Poirot guides us through a staggering array of red herrings and false leads, helped and hindered by the faithful Captain Hastings.

Strychnine would become a popular choice for Christie, and it makes an appearance in four further novels and five short stories, killing a total of five characters. Christie really makes the most of this poison in *The Mysterious Affair at Styles*, and she takes every opportunity to show off her extensive chemical knowledge.

The strychnine story
Strychnine is a plant alkaloid, an odourless solid with a very bitter taste. It forms long, thin, colourless crystals that are poorly soluble in water (1g dissolves in almost 6.5 litres). The compound is obtained from the plants of the genus *Strychnos*. There are many *Strychnos* species, which grow in the warm regions of Africa, America and Asia, and several of them contain strychnine, but the compound is most abundant in *Strychnos nux-vomica*, a tree native to India. Strychnine is found in the large, disc-like seeds of the plant,* and it can be extracted relatively easily.

In regions where the plants are native, the poisonous properties of strychnine have been known for centuries, and extracts of the plants have been used as pesticides. In India strychnine is still used in 'hudar' capsules to elevate blood pressure, whereby the seeds are soaked in water or milk beforehand to reduce the toxicity; taking strychnine might put your body under sufficient stress to raise blood pressure, but this is a consequence of the symptoms, rather than a

---

*Another poison, brucine, is found in the bark of the shrub.

direct result of the action of the compound (there are other, more reliable, ways of controlling blood pressure). *Strychnos nux-vomica* also crops up in homeopathic remedies today, though the extremely high dilutions used in homeopathy mean that anyone taking these remedies is unlikely to be poisoned. Strychnine's use in conventional medicine was on the wane by the time that Christie wrote *The Mysterious Affair at Styles* in 1916, and by a few years later it had disappeared completely.

Strychnine can be inhaled or injected but it was normally administered orally, whether for criminal or medical intent. To improve the solubility of strychnine in water it was usually converted to a salt. This does not affect the toxicity of the compound, but it does make it easier to administer. When swallowed, strychnine salt is not absorbed from the acidic environment of the stomach, but is readily taken into the body through the walls of the small intestines. Once it has passed into the bloodstream strychnine can be easily distributed throughout the body. The target site for strychnine is, as so often in toxicology, a receptor in the nervous system.

How strychnine kills

Strychnine acts on the central nervous system (CNS), the network of nerve cells that send, receive and coordinate messages around the body. Signals that control voluntary movements are sent along motor neurons. These are cells that form a fine strand, the axon, just a few micrometres (a millionth of a metre) across. An axon extends out from the main body of the cell carrying electrical signals, and can be more than a metre long. Motor neurons stretch from the spinal cord out to the extremities of our body. The junctions between motor neurons are known as synapses. Chemical signals, or neurotransmitters, are released from the end of one neuron; they cross the gap and bind to a receptor on the next neuron, triggering a pulse of electric charge to transmit the signal further. The signal will terminate at a muscle, where neurotransmitters cross a synapse to cause the contraction of the muscle fibre.

At rest, the inside of a nerve cell is more negatively charged than the outside; it is a switch in this polarity that is the source of the electrical signal passing down the axon. The neurotransmitter that crosses the synapse to trigger the electrical pulse in the connecting nerve is acetylcholine; another chemical, glycine, is released to counteract this effect. When glycine binds to receptors on a nerve it increases the flow of negative charge into the cell, making it more difficult for the nerve to produce a signal. This is an important control measure (like a system of brakes), ensuring that the nerve does not fire at the slightest provocation.

Strychnine binds to the glycine receptor three times more effectively than glycine does. By blocking the glycine receptors, the moderating effect of glycine is effectively switched off; the nerve will now fire at the smallest trigger (the brakes are disabled, and the system careers out of control). The muscles

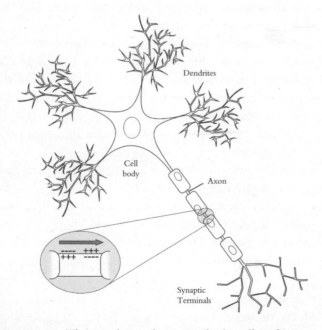

A motor neuron. The switch is polarity inside the cell is the cause of the electrical signal that passes along the length of the axon.

connected to the motor neurons will contract fully, and for extended periods, at the slightest stimulation.

In humans the muscles of the back tend to be stronger than those on the front of the body; strychnine causes its victims to arch their backs, with the whole body resting only on the back of the head and the heels. In some cases the muscle spasms affect the stomach muscles and throw the victim forwards into a curled position or, if the side muscles are affected, rapid side-to-side convulsions can be seen. When *all* the muscles are affected, the arms are held rigidly at the side of the body, with the muscles in the face contorted to give the victim a sardonic grin, and the eyes bulging and flitting from side to side.

Although the primary effects of strychnine are on the motor neurons, the compound also affects the neurons in the cortex of the brain. The victim is conscious throughout; their senses and awareness of their situation are heightened by the strychnine increasing the sensitivity of the nerve endings in the brain. In the roster of truly appalling poisons, strychnine must rank close to the top of the list.

Symptoms appear between 15 and 30 minutes after strychnine's ingestion. They start with a tingling and twitching in the muscles. There may also be nausea and vomiting. As the symptoms progress, the muscle spasms become more violent, and eventually waves of muscle spasms bring the whole body into convulsion. Attacks of convulsions are interspersed with periods of relative calm. Victims often have a very red complexion because their muscles are working so hard, burning up oxygen at an incredible rate. For the victim this is exhausting, and the sufferer rarely survives more than five such attacks, with death occurring between one and three hours after ingestion of the poison. Death is caused by asphyxiation, as the muscles controlling breathing are affected.

Is there an antidote?
The effects of strychnine on the body are so dramatic that death usually occurs rapidly, so any treatment has to be given quickly. There is no specific antidote for strychnine, but there

are ways of alleviating the symptoms. Morphine can be given for pain relief and sedation, but it is most important to treat the convulsions. Muscle relaxants will stop these, allowing the body to process and excrete strychnine without adverse effects.* Today there is a choice of anti-convulsion drugs, because of their application in surgical procedures as well as in emergency treatment. Diazepam (also known as Valium) would be the preferred choice when treating strychnine poisoning, with the patient being supported with artificial respiration to maintain breathing. Diazepam was not available until 1963, so the choices available in 1920 would have been more limited. Other barbiturates would have been accessible, though, and these would have been the recommended treatment for strychnine poisoning at the time.

There is one compound that is a very effective muscle relaxant, and its ability to counteract the effects of strychnine has been known since 1850. Curiously, it is found in a plant of the same genus, *Strychnos*. The infamous arrow poison, curare, is an extract of *Strychnos toxifera*. It contains the alkaloid *d*-tubocurarine (so named because it was extracted from a tube used to store arrow poison). *d*-tubocurarine works in a similar way to strychnine, by blocking neurotransmitter receptors in motor neurons, but in this case the neurotransmitter in question is acetylcholine rather than glycine. The binding of acetylcholine to its receptor is the signal for the nerve to fire. When the acetylcholine receptor is blocked the nerve cannot fire, and the muscles relax (so this is the direct opposite of strychnine's blocking of glycine receptors). An account of the use of *d*-tubocurarine in anaesthetic practice was first published

---

*Steps can also be taken to prevent convulsions by keeping the patient calm and still in a dark and quiet room. Without stimuli, the nerves will not send signals, allowing time for the strychnine to be eliminated from the body.

in 1942, though it did not become standard practice in Britain until the 1950s.

Activated charcoal can also be administered to prevent further absorption of strychnine into the body. Administering activated charcoal is now a standard procedure when dealing with patients who have overdosed on a wide range of compounds, but the procedure has been recognised since the 1830s. In 1831, in a dramatic demonstration in front of his colleagues at the French Academy of Medicine, the pharmacist P. F. Touery swallowed ten times the lethal dose of strychnine mixed with charcoal. He and his colleagues waited several hours for signs of poisoning to develop but none were observed.*

Some real-life cases

Strychnine has been used many times in poisonings; it is listed third in the top ten poisons by number of criminal cases, behind only arsenic and cyanide. It has been used in some highly creative ways, with cunning often required to disguise the taste or to ensure that the victim swallows the lethal dose in one gulp.

One real-life case of murder with strychnine bears some striking resemblances to the method used in *The Mysterious Affair at Styles*. This happened in 1924, four years after the publication of the novel. It involved a wireless operator, Jean-Pierre Vaquier, who had travelled to England in pursuit of Mrs Mabel Jones, with whom he had fallen desperately in love. The couple had met and begun an affair in France, where Mrs Jones was convalescing after an illness. Mabel returned to England and to her husband, Alfred, the proprietor of the Blue Anchor Hotel in Byfleet, Surrey, who had stayed in England while his wife was recuperating.

When Vaquier arrived in England he went to stay at the Blue Anchor Hotel, and the affair continued under Alfred's

---

*Yes, we have seen almost exactly the same story before – a different Frenchman, on a different date, with a different poison, arsenic (see page 40). The French Academy of Medicine clearly needed some persuading as to the merits of activated charcoal.

roof. Alfred was in the habit of drinking a little more than was good for him, so he regularly took bromide powders to counteract the effects of alcohol. He would take a dose of these powders from a blue bottle kept on a shelf in the hotel bar each morning, and dissolve it in a glass of water.* One morning he added a dose from the blue bottle to a glass of water as normal, but he remarked that the powders did not fizz as they normally did. He drank down the mixture in one gulp regardless, and exclaimed how bitter it was.

Mabel Jones looked in the blue bottle and saw some long crystals amongst the usual powder. She tasted the crystals, and noticed the bitter flavour. She gave her husband some salt water to make him sick, and some tea with soda to try to counteract the effects of whatever it was he had swallowed. It was to no avail. Soon after, Alfred complained of numbness and of feeling cold. He went to bed and a doctor was called. By the time the doctor had arrived Alfred was in the midst of agonising convulsions. At 11.30 a.m., an hour and a half after drinking the poison, he was dead.

The circumstantial evidence against Vaquier was pretty strong. He certainly had a motive to dispose of Alfred, and he also had an opportunity, although no one saw him add strychnine to the blue bottle of bromide powders. When the bottle was recovered by police it had been washed out, but tests on the residual water inside the bottle confirmed the presence of strychnine. Vaquier had purchased strychnine, among other compounds, under the pretext of using it for wireless experiments. He had signed the poison register, but used a false name. A wireless expert was called to testify at the trial, and he confirmed that there were no known applications of strychnine in this field. Vaquier was found guilty and hanged; it is not recorded if Jean-Pierre Vaquier was a fan of Agatha Christie novels.

Christie cannot be blamed for the popularity of strychnine in murders anyway. There had been plenty of cases to inspire

*You can probably see where this is going.

would-be poisoners and murder-mystery writers before *The Mysterious Affair at Styles* was published. The most famous of these is probably that of Dr Thomas Neill Cream, a medical doctor who, in 1892, murdered four women. Cream had already served a sentence for one murder and was a suspect in the investigation into another, with both of these occurring in the United States. On his release he travelled to England, and within weeks of arriving in London began poisoning prostitutes. After meeting each of his victims he persuaded them to swallow a pill, which he said would improve their complexion. The pills in fact contained a lethal dose of strychnine, and the women died in agony hours later.

Cream was caught; it took only 12 minutes for the jury to reach a guilty verdict, and he was sentenced to hang at Newgate prison. There were whisperings that a man standing next to Cream on the scaffold heard his last words as the bolt was pulled from the trapdoor: 'I am Jack the …'*

Agatha and strychnine
The Mysterious Affair at Styles provides such an insight into the chemistry of strychnine that it is worth looking at this novel in some detail. In the early hours of the morning the Styles household rushes to the aid of the very unwell Mrs Inglethorp. As they burst into her bedroom, they witness the old lady in the throes of the most violent convulsions. The seizures rack her body with such force that she has managed to knock over a heavy bedside table. A second bout of convulsions forces her body to arch in such a way as to leave her resting on the back of her head and her heels, with her belly forced up towards the

---

*The noose is supposed to have choked off the end of the sentence. But sadly this is not the solution to the Jack the Ripper case, as Cream was in prison at the time that one of the murders took place. It could represent a bit of showmanship on his part – he seemed to crave notoriety. More likely, the story was made up after the event, as there was no mention of it at the time. Perhaps the hangman invented the story, to claim the credit for hanging this infamous murderer.

ceiling. Unsuccessful attempts are made to administer brandy to the victim (though what they hope to achieve with this is unclear). With her last gasps of breath, Mrs Inglethorp cries out the name of her husband before collapsing back on the bed, dead. Resuscitation is attempted, but to no avail.

The signs of strychnine poisoning are characteristic, and even the rather obtuse Captain Hastings recognises the symptoms immediately. Christie uses strychnine to kill off a total of five characters in her books, and only once is death suspected as being caused by anything other than strychnine – in *Poirot Investigates*, where a death is attributed to tetanus. The convulsions can be similar; this is where the word 'tetanic' comes from in describing the type of convulsions experienced by Mrs Inglethorp. 'Grand mal' seizures of epilepsy patients can appear very similar to strychnine poisoning, but Mrs Inglethorp was not epileptic and this scenario is never considered.

The death of Mrs Inglethorp is clearly suspicious, and a post-mortem is ordered, with the results confirming poisoning by strychnine; some of the poison is found in her stomach, with more found in other body tissues. The total dose is estimated to have been just under a grain (approximately 65mg). This is a little low for a lethal dose, which is normally quoted as 100mg, but 65mg may well have been enough to dispatch an older female victim. Indeed, a dose as low as 36mg has proven fatal.

Rigor mortis sets in quickly after strychnine poisoning, and this has been known to lock the corpse in the pose of its final convulsion. This does not occur in all cases, and very large doses of strychnine can cause a rapid death with no convulsions at all. *Rigor mortis* wears off in time and strychnine would leave no obvious sign of internal damage in the body, but being aware of the circumstances of the death, a pathologist or toxicologist would know to look for the poison in the stomach contents. Strychnine could be extracted from the body using the Stas method (see page 169). Specific and reliable chemical tests for strychnine were well established by 1920, but before this pathologists would have recognised the distinctively bitter taste of the poison. Any doubts could be assuaged by feeding the unknown poison to an animal and comparing the symptoms to an animal fed with strychnine.

Today the extraction and identification of poisons is much more straightforward, of course. A range of analytical techniques is available to confirm the presence and identity of a poison. The main limitation on modern methods is making sure the right tests are carried out for the appropriate poison; there are so many compounds that can prove fatal if taken in the right dose that toxicological screenings normally look only for the usual suspects. Though no longer specifically tested for, strychnine would be detected as part of a regular toxicological screening for alkaloids at post-mortem.

Strychnine's heyday as a murder weapon is long gone, thankfully, owing to tighter controls on its export and use. Strychnine is banned in Britain, though it appears to remain frighteningly easy to obtain dangerous quantities in the United States. Between 1949 and 1979 there was approximately one death a year in Britain from strychnine poisoning, and this was usually self-administered and obtained from stocks kept for agricultural purposes. Occasionally strychnine cases present themselves in hospitals even today, but this is often due to an overdose of hudar traditional Indian remedies, with the symptoms being twitching or convulsions that tend not to lead to death.

The cause of Mrs Inglethorp's death would have been easy to establish. What was much more difficult to figure out was *how* it was done. First the poisoner had to get hold of strychnine. Today the sale and use of strychnine are carefully controlled, but things were a little more relaxed in 1920. Hospital dispensaries would stock strychnine, as it was prescribed as a stimulant. Pharmacists would keep it for sale as a pesticide or for use in tonics. Although strychnine was known to be highly toxic and the sale of it was restricted, this imposed few obstacles to the determined poisoner.

As is the case with many other poisonous substances, the purchaser had to be known to the chemist and have a valid reason for the purchase before the sale could proceed. Both the buyer and the seller would have to sign a poison register recording the amount and intended use of the poison, along with the name and address of the purchaser. At least the poison register gave some measure of traceability in theory, but some poisoners – rather unsportingly – have been known to use false names and disguise their handwriting.

Having strychnine in your house today would be suspicious, but it would not have been unusual back in 1920. In *The Mysterious Affair at Styles* a bottle of strychnine is found in the hospital dispensary, from where several suspects had the opportunity to steal some. Another bottle of strychnine is found in a drawer in the house. Strychnine is bought from the local pharmacist to poison a dog (with the poison register signed in accordance with the law), and finally, there is strychnine in Mrs Inglethorp's tonic, a bedside drink to pep her up. None of these is in itself suspicious but, as Poirot puts it, '... there is altogether too much strychnine about this case'.

Strychnine is unremarkable in appearance, and at a passing glance it's indistinguishable from salt or sugar. A closer inspection may reveal that the long crystals of strychnine are different from the small blocks of salt or sugar crystals. It would be fairly easy, then, to add some colourless crystals to food or drink without being suspected, so long as no one was paying too much attention. However, strychnine is one of the bitterest substances known, and can be detected in water in quantities as low as one part in 70,000. This means that the murderer would have to dilute a fatal dose (100mg) in seven litres of water to disguise the taste, so the victim would be likely to be alerted to its presence pretty quickly, or they would be suspicious that they were being made to drink so much. Either the flavour has to be disguised or concealed somehow to prevent the victim from becoming suspicious, or the fatal dose has to be swallowed in one gulp. Four possibilities are examined in the novel: the evening meal; a cup of coffee taken just before bed; in cocoa kept in a warming pan by the bedside of the victim; or in one of the medicines taken by Mrs Inglethorp.

First, the evening meal. Mrs Inglethorp ate the same as everyone else, but she did not eat very much. To avoid the bitter taste a fatal dose of strychnine would have to be thinly distributed throughout the food, so it seems unlikely that the poison could have been administered this way. Also, the death occurred many hours after the meal was eaten, and it usually takes less than 30 minutes for the effects to manifest themselves. The onset of symptoms might be delayed slightly by a heavy meal, but that does not seem possible here.

Second, the cup of coffee. This would seem a very good contender, as the strong, bitter taste of the coffee could cover the taste of the strychnine. There is a difficulty, though, because the coffee was taken after the evening meal, and the symptoms did not present themselves until the early hours of the morning. Other drugs could have been added to delay the onset of symptoms, but more of this later. In the end the coffee is eliminated, as Poirot discovers a fresh coffee stain on the carpet where Mrs Inglethorp must have dropped the cup, so she never actually drank it.

Third, the cocoa. This can be dismissed almost immediately, as it would not be possible to disguise the bitter taste of strychnine, and it would be unlikely that anyone drinking a strychnine-laced cup of cocoa would take more than a sip. However, Poirot determines that there is another substance dissolved in the cocoa, a narcotic. The word 'narcotic', from the Greek meaning 'to make numb', can mean different things to different people, depending on whether you are a layperson, work in law enforcement or work in the medical profession. We are never told what the narcotic is or what Christie means by the term, but we can assume that she was probably referring to a morphine-like substance because of the effects it produces in the victim.

In addition to its well-known pain-relieving effects, morphine is very good at inducing sleep. It is unlikely that morphine would induce dramatic tetanic seizures so it was not a morphine overdose that did away with Mrs Inglethorp, but morphine *was* involved in her demise.

Morphine affects the digestive tract by impairing the muscle contractions that normally move food from the stomach to the small intestines and on through the many metres of gut. For years it was used as a treatment for diarrhoea for this very reason, and it is why many users of morphine-like drugs complain of constipation. This has the effect of delaying the movement of food from the stomach to the small intestines by as much as 12 hours. As we have seen, strychnine is not absorbed in the stomach owing to the acidic environment, but it is absorbed through the small intestine. This could explain the delayed effects of the strychnine dose taken by Mrs Inglethorp, several hours before the convulsions started.

Finally, we must look into Mrs Inglethorp's medicine cabinet. Here there are two possibilities: the sedative powders and the tonic. The small packets of sedative powders contained potassium bromide. Bromide powders could be purchased in boxes containing several individual doses from pharmacists. Each dose would be a few grams of white powder folded up in paper. The powder would be dissolved in water and swallowed. Potassium bromide was widely used as a sedative for epilepsy, as it calmed the seizures, and more extensively as a general sedative. It was commonly thought that bromide was added to soldiers' tea during the Second World War to reduce libido. This is unlikely to be true – bromide powders would also act as a sedative and reduce alertness, a distinctly undesirable side effect in an army at war.

Bromide powders are not completely innocuous, and with a high level of use can cause serious health problems. Bromide is retained in the body for a long time. The half-life for elimination is nine to twelve days, so it is relatively easy to ingest more than is being excreted if the powders are taken regularly over an extended period of time. Could bromide poisoning, known as bromism, account for the symptoms displayed by Mrs Inglethorp on the fatal night?

People with bromism display a wide range of symptoms including lethargy, slurred speech, headaches and psychiatric effects such as depression and confusion. Seizures have been observed in some cases, but not of the type seen in strychnine poisoning, and death from bromism is rare. Any symptoms would have been present a long time before Mrs Inglethorp's death, so the bromide powders can be discounted. There is, of course, the possibility that the powders had been tampered with and strychnine added, but the last powder had been taken two days previously, so again they can be discounted.

Mrs Inglethorp also took a tonic every night. This was a pick-me-up or stimulant quite commonly used in the early part of the twentieth century; the principal ingredient was strychnine. It was believed to stimulate digestion and make you feel brighter and more alert. A most unusual example of its use occurred in 1904, when athlete Thomas Hicks won the marathon at the St Louis Olympics. During the gruelling race Hicks was given two doses of 1/60 grain (approximately 1mg) of strychnine and at least a flask of brandy by his trainers, in the belief that it would keep him going. He managed to finish but he had to be carried across the line, and was too weak to collect his medal. If Hicks's trainers had been any more helpful they could have killed him.

Research has debunked the claims that strychnine acts as a stimulant. Doubts over strychnine's efficacy started in the 1950s, and by 1972 it was completely discredited as a therapeutic agent. In 1920, though, small quantities would be diluted to 'safe' levels and sold over the counter without prescription. Mrs Inglethorp would not be unusual in having a bottle of strychnine-based tonic on her bedside table. The only problem with the tonic as a source of the poison is that Mrs Inglethorp had swallowed the last dose on the night she died. Even taking several spoonfuls at once would not be fatal. If she had been taking the tonic previously with no ill effects then clearly there could not have been a mistake in making up the prescription.

The question of chronic strychnine poisoning is raised at the inquest. Could the levels of strychnine have built up in Mrs Inglethorp's body over several days, resulting in a final, fatal dose? Chronic strychnine poisoning is unlikely; the compound is eliminated from the body fairly rapidly, either unchanged or modified by enzymes in the liver, with a half-life of about ten hours. However, anyone with liver disease such as hepatitis or chronic alcoholism would be more susceptible because their liver would not be working efficiently, and the drug might accumulate in the organ. Mrs Inglethorp was in good health, but anyone suffering from the combined effects of liver disease and chronic strychnine poisoning would be likely to display symptoms such as twitching or tremors. If an error had been made when the tonic was made up and this had resulted in a stronger solution, it would have produced clear symptoms in the victim rather than one sudden, dramatic and fatal attack.

It was not one drug that did away with Mrs Inglethorp; it was a combination of three compounds and a lot of planning. Of all the possible sources of strychnine to be found in the house, it was Mrs Inglethorp's own tonic that caused her death. The whole bottle of tonic contained enough strychnine to kill, but it all had to be concentrated into the final dose. This was actually simple to achieve.

Strychnine is poorly soluble in water and is normally converted into a salt form to increase its solubility, but the choice of salt is important as not all strychnine salts dissolve well in water. Strychnine sulfate was used to make up the tonic, and this would dissolve perfectly well to leave a clear, colourless solution with the strychnine distributed evenly throughout the bottle. A problem occurs if other salts are added to the mix. Depending on the salt used, it can cause the strychnine salt to convert from a soluble to an insoluble form. Potassium bromide added to a solution of strychnine sulfate would cause, over the course of a few hours, the formation of colourless crystals of strychnine bromide, which would settle to the bottom of the bottle.

The potassium bromide needed to precipitate the strychnine was obtained from Mrs Inglethorp's bromide powders. One or two powders would have been added to the tonic when it was purchased. This would have been more than enough to create the desired crystallisation of strychnine bromide. The addition of the powders to the tonic would not have changed its appearance or taste. Potassium bromide dissolves in water easily to form a colourless solution and has only a slightly sweet taste when dilute but a bitter taste at higher concentration. If the person who administered the tonic did not shake the bottle before pouring out Mrs Inglethorp's regular dose, the victim would have swallowed almost the entire quantity of strychnine in one go. Addition of morphine, or another narcotic, to the cocoa would delay the onset of strychnine poisoning and divert attention from the tonic, which Mrs Inglethorp took just before going to bed.

To explain this highly devious method to Captain Hastings, Poirot reads from a book on dispensing he found at the hospital dispensary.

The following prescription has become famous in text books:

Strychnininae Sulph	*gr. I*
Potass Bromide	*3vi*
Aqua ad	*3viii*
*Fiat Mistura**	

This solution deposits in a few hours the greater part of the strychnine salt as an insoluble bromide in transparent crystals. A lady in England lost her life by taking a similar mixture: the precipitated strychnine collected at the bottom and in taking the last dose she swallowed nearly all of it!

*'Fiat mistura' roughly translates as 'becomes a mixture'.

The above text is taken from *The Art of Dispensing*, which Christie would have studied for her exams when she qualified as a dispenser.

The descriptions by Agatha Christie of strychnine and its poisonous effects are very accurate in all the stories where she uses it. The chemical knowledge needed to plot a murder mystery such as *The Mysterious Affair at Styles* is considerable, and as I mentioned earlier, this was remarked upon at the time in a review of the novel in the *Pharmaceutical Journal and Pharmacist*. The academic journal praised her scientific accuracy in the novel; no wonder she was so proud of this review.

THALLIUM

The Pale Horse

I looked and there before me was a pale horse! Its rider was named Death, and Hell was following close behind him.

Revelations 6:8

The 'Pale Horse' of Agatha Christie's novel was an organisation that arranged deaths on demand. At the Pale Horse Inn, a witches' coven of contract killers appears to commit murder for money. None of the so-called witches meets any of the victims, all of whom seem to die of natural causes. Paranormal theories abound, but there is a much more down-to-earth explanation. Mark Easterbrook, an historian and writer, is drawn into the mystery when he witnesses a fight between two girls in a coffee shop, one of whom has a clump of hair pulled from her head. A week later he sees that one of the girls from the fight has died, and this is just one from a list of unexpected deaths. A series of unusual events leads Easterbrook to suspect foul play, and he sets out to prove that the deaths are connected and that someone, somewhere, is responsible for them. It turns out that thallium has been given to all of the victims in low, regular doses, which accumulate in the body and finally kill after days, or even weeks, of pain and suffering.

Thallium is sometimes known as the 'poisoner's poison'. It was little known prior to the publication of *The Pale Horse* in 1961, and so was rarely tested for in post-mortems. The idea of using this element in a murder plot was suggested to Christie by an American doctor, but she must have carried out considerable independent research to develop her detailed knowledge of this rarely used poison. Thallium poisoning can result in a wide range of symptoms that are easily attributed to many different natural diseases. Despite the obvious benefits of thallium to the murder-mystery writer, Agatha Christie uses it in just one novel. What she lacks in the number of books, though, she more than makes up for in the number of deaths. A total of ten victims are named in *The Pale Horse*, with more murders implied.

This book has gained some notoriety, and it has even been suggested that it was the inspiration for several real-life murders. But from another point of view, Christie's writing about the symptoms of thallium poisoning in such a prominent and accurate way raised awareness of this deadly element, and may even have saved lives.

The Thallium story

Thallium (Tl) is the 81st element in the periodic table; in its pure state it is a soft grey metal. It was indirectly (and independently) discovered in 1861 by William Crookes (1832–1919) and Claude-Auguste Lamy (1820–1878). They had been analysing different materials by burning them and observing the colour of the flame produced, using a recently invented technique called flame spectroscopy. This method splits light up into its component colours. Each element in the periodic table burns with a characteristic colour, so different elements present in a sample can be identified using a spectrometer. Flame spectroscopy is still in use today; a version of this technique known as atomic absorption spectroscopy can be used to analyse tissue and fluids from a body at a post-mortem to identify poisonous elements such as arsenic, mercury and thallium.

In the samples analysed by Crookes and Lamy, a flame colour was observed in the 'green' part of the spectrum, where no element had previously been observed, so they knew there must have been an unknown element present. The green colour led Crookes to name it 'thallium', after the Greek *thalos*, meaning 'green shoot or twig'. Crookes continued his experiments on the new element, and announced his discovery in the March 1861 issue of *Chemical News*. Lamy announced *his* discovery some months later, and an almighty row ensued over who had prior claim to the discovery. To keep the peace, credit was given to both scientists. However, what both Crookes and Lamy had actually discovered was a *compound* of thallium, and not the pure element. The world had to wait another year for Lamy to isolate the element and produce a tiny ingot of pure thallium.

Thallium is relatively abundant on Earth but is only thinly distributed in rocks and soils, because it reacts readily to form compounds that are often very soluble. Water will erode away thallium salts and deposit them elsewhere in an endless process of redistribution. This means that thallium, despite its toxicity, is not considered to be an environmental pollutant as people, livestock and crops are highly unlikely to be exposed to dangerous levels. Unlike thallium salts, which have no odour, very little taste and are highly toxic, thallium metal is unlikely to poison anyone, because it is not soluble in water and therefore cannot be transported into the body easily.*

---

*There are two ions of thallium, Tl^+ and $Tl3^+$. Tl^+ is the most common, resembling the ions of the alkali metals in group 1 of the periodic table in its chemistry, and potassium (K^+) in particular. Tl^+ will form strong chemical bonds to negatively charged atoms or groups of atoms to form a salt, such as thallium chloride (TlCl).

Besides their potential as poisons, thallium salts have several applications, though their use is carefully monitored and controlled today. Thallium oxide and some other thallium compounds are added to specialist glass to increase the refractive index (the degree to which light bends when it passes through a different material). These specialist glasses are used in the optics industry, to make camera lenses, for example. Once in the glass the thallium compound is trapped, and completely safe as it cannot leach out. Other modern applications include uses in components in the electronics industry.

Before their toxicity was properly realised, thallium salts had applications in medicine. One effect of administering thallium salts was discovered by accident in the 1890s, when thallium acetate ($TlCH_3CO_2$) was given to tuberculosis patients in the hope that it would reduce the night sweats they experienced. It had no effect on these at all, but the patients' hair fell out. Hair loss today would be considered a symptom of damage to the body, but at the time this was seen as a potential benefit, particularly for the treatment of ringworm.

Ringworm is a fungal infection of the skin, and to treat it effectively it is best to remove the hair from the affected area. By the 1920s the standard treatment was 8mg per kilogram body weight of thallium acetate, given in a single dose. By around 15 days later all the hair on the head would have fallen out, exposing the area affected by ringworm. Sulfur drugs were then administered daily to treat the fungus. This method was considered safe, despite a growing appreciation of the toxic effects of larger doses of thallium salts. Around 40 per cent of those treated for ringworm reported mild side effects, and 25 per cent suffered more severe problems, with pains in the legs and stomach upsets. The alternative method of treatment was to use X-rays to make the hair fall out; by comparison, thallium was probably the lesser of two evils.

Later, it was realised that the patient didn't have to swallow the thallium acetate, as it could be applied as an ointment to the affected area. Thallium would then be absorbed through the skin with the same effect, but only a localised patch of hair would fall out. By the 1930s thallium acetate had become such a standard in the treatment of ringworm that it was sold without prescription in over-the-counter remedies called Celio and Koremlou. A 10g tube of one of these typically contained 700mg of thallium acetate.*

In 1930s pharmacies, alongside creams for ringworm, another thallium salt would have been on sale, only this one was a pesticide; somehow, no one thought it odd that thallium compounds were considered completely safe for humans, yet deadly to other animals. Thallium sulfate was added to a sugary syrup that was particularly attractive and effective for rats, cockroaches and ants. Accidents, suicides and even murders using thallium pesticides led to it being banned in the United States in 1972, with many other countries rapidly following suit.

Alternative methods of treating ringworm are now available, and pesticides contain compounds that are more toxic to the target pest than to humans, so you won't find thallium salts for sale in pharmacies today. But there is one medical application of thallium still in use: the thallium stress-test. A radioactive form of thallium, known as thallium-210, is injected in sub-lethal doses (as the chloride salt). The radioactivity emitted by the thallium-210 is detectable from outside the body, and is used to monitor the health of the heart when the patient does moderate exercise. This technique allows specialists to monitor blood flow in the hearts of patients with coronary heart disease, as the thallium-210 is only taken up by healthy parts of the heart muscle, where there is a good blood flow.

*These were used in several attempted suicides. Thankfully most people only swallowed one tube, which was not quite enough to kill, but it would have made them very ill.

How thallium kills

Precisely how thallium interacts with the human body is not fully known. It has no natural biological role, but because of its similarity to potassium it is readily absorbed into the body. Potassium is very abundant in the human body; there is approximately 120g of potassium in the average 70kg man, and it performs a variety of roles. Thallium can replace potassium at all sites in the body, but it will not perform the same functions, leading to a degeneration of all the processes that normally involve potassium. The symptoms of thallium poisoning are a direct result of potassium-based process malfunction.

One of the most important potassium roles is in nerve function. Uptake of thallium in nerve cells is high, because of the abundance of potassium needed by them to generate an electrical signal. Once inside the cell thallium causes significant damage (especially to the long part of the cell, the axon).

Other biological functions of potassium include the release of adenosine triphosphate (ATP), the body's source of energy. Cells with high energy demands, such as nerves, heart and hair follicles (the latter due to their being replaced at a high rate) are therefore particularly badly affected when thallium replaces potassium. There are more problems inside the cell: potassium is important for stabilising ribosomes, structures that build proteins that enable the body to grow, repair itself and carry out chemical reactions. Another important potassium role lies in the production of thiamine, a B vitamin; thallium poisoning can resemble a deficiency of this vitamin.

Thallium has a particular affinity for sulfhydryl groups (–SH), which are common in proteins. It binds to these groups, and disrupts the protein's function. Thallium doesn't target any one protein or enzyme in particular, which explains the wide range of symptoms observed in poison victims. The strength of the interaction between thallium and the sulfhydryl group is not strong enough to trap the thallium permanently at a specific site, so it can dissociate and move on to other proteins or structures in the body, causing further chaos.

The majority of these effects are reversible once thallium has been excreted from the body. It can be excreted in urine, faeces, saliva and sweat, and can also be lost through breast milk and tears. Opinion on how long it takes to excrete thallium via these bodily functions varies, but the size of the initial dose is important. At low doses the half-life of excretion may be between one and three days; larger doses can take a month or more for half the amount of thallium to be lost. Because thallium does not perform the same chemical reactions as potassium, despite its ions being a similar size and identical in charge, cells can recognise the difference, and the body will attempt to excrete thallium, passing it into the gastrointestinal tract. The problem is that as the thallium travels further along the tract it is reabsorbed, because of this likeness to potassium. In this way the same dose of thallium can keep re-poisoning and damaging the body in a cycle of excretion and reabsorption across the walls of the whole gastrointestinal tract.

Owing to its long half-life in the body, levels of thallium can accumulate rapidly through regular small doses until a lethal level is attained. This process causes chronic symptoms over the period of poisoning. No effects may be seen for up to 24 hours, but there may then be symptoms similar to those of mild flu. Thallium salts are irritants, so these initial effects include nausea, vomiting and diarrhoea, but these subside, to be replaced by severe abdominal pains. As the days go by and the doses of thallium accumulate, the symptoms grow progressively worse. There may be muscle weakness and atrophy, a tingling and numbness in the extremities; damage to the peripheral nervous system, and painful legs, with the feet feeling as if they are on fire; the body becomes very sensitive to touch. There can be psychological effects, too, with mood swings, sleeplessness, periods of confusion and even hallucinations.

About two weeks after first ingesting thallium salts, the hair starts to fall out – a classic signal of thallium poisoning, one that can leave the victim completely bald. There may also be skin-pigmentation changes and irritation, particularly around the roots of the hair, due to the effects of the poison on sweat glands. Around three weeks later Mees lines – white horizontal lines – appear on the fingernails and toenails. These lines are also seen in arsenic poisoning (see page 38), but they are less marked with that poison than in thallium poisoning. Unlike in arsenic poisoning, thallium is not reliably retained in the hair,[*] despite its affinity for sulfhydryl units (which are abundant in hair and nails).

An alternative murder method to gradually poisoning the victim is to give them a single fatal dose. For humans this is 12–15mg for every kilogram of body weight; approximately 1g would constitute a fatal dose for an adult. The symptoms of acute thallium poisoning appear rapidly, with vomiting and diarrhoea occurring within hours, and severe neurological symptoms appearing between two and five days later. These symptoms can include abdominal pain, nausea, vomiting and diarrhoea, dramatic weight loss (owing to the vomiting the body cannot absorb enough nutrients to maintain body weight), delirium, slower breathing, and in a short space of time seizures, coma and death, which can occur any time between hours and weeks after the thallium is administered, depending on the dose and what, if any, treatment is given.

Is there an antidote?
When *The Pale Horse* was written in 1961, there was no standard treatment for thallium poisoning other than supportive care until the body could excrete the poison of its own accord, with luck. If thallium poisoning was suspected (which it rarely was), the suggested method of treatment was as follows: 1) removal of ingested thallium by stomach-pump, followed by treatment with activated charcoal; 2) promotion of urinary

---

[*]Some might be retained, but not enough to determine exposure to the metal reliably – this again is due to thallium's weak bonding to sulfhydryl groups.

excretion by abundant intake of fluids, followed by the administration of potassium chloride. If the thallium had been absorbed through the skin, then only the second part of the treatment was liable to be of any use.

A few years after the publication of *The Pale Horse*, research began on finding an antidote for thallium poisoning. Initial attempts used dimercaprol, or British Anti-Lewisite, a chemical developed during the First World War as an antidote to Lewisite, an arsenic-based chemical weapon. Dimercaprol contains sulfhydryl groups for which some metals, especially arsenic, have a high affinity. The compound binds to arsenic and other toxic metals, allowing them to be excreted safely. Unfortunately, dimercaprol has almost no effect on thallium. Even though the metal has affinity to sulfhydryl groups, it does not bind with sufficient strength to make dimercaprol an effective treatment for thallium poisoning.

A more successful antidote was dithizone, which significantly increased the amount of thallium excreted in urine but was not without its problems. Dithizone, though better than dimercaprol, still does not bind strongly enough to thallium to prevent some of it being re-released into the body; dithizone and other chemicals that work in this way (known as chelating agents) can therefore take thallium from areas where it has been sequestered and is causing few problems, and reintroduce it into a more active and dangerous role within the body. In other words, chelating agents can make poisoning symptoms worse, even while the patient is being treated.

Thallium can also be removed from the blood using dialysis, a process that forces thallium to diffuse out of the blood through a membrane. This process needs to be repeated again and again as more thallium leaches out of tissues and organs into the bloodstream. The leaching process can be speeded up by giving the patient potassium chloride. This helps displace thallium from around the body. However, Prussian Blue – a compound we have met a number of times (see pages 72 and 77) – is nowadays the treatment of choice for thallium poisoning because it binds to the metal more effectively than other agents, and has no known hazards associated with it. Doses of 250mg per kilogram of body weight are administered orally to trap the thallium, and to prevent its absorption into the body. The thallium-bound Prussian Blue is then excreted via the bowels, often turning the faeces blue. This treatment was established in the early 1970s – too late for the victims in *The Pale Horse*.

Some real-life cases
The first known case of murder by thallium poisoning in Britain occurred in 1962, only months after *The Pale Horse* was published. A coincidence? Many thought not, but then many things are more obvious in hindsight. The murder in question was of Molly Young by her 15-year-old stepson, Graham. Graham Young had been obsessed by the macabre, and poisons in particular, since he was a small boy, but the turning point in his career as a prolific poisoner was when his father gave him a chemistry set as a reward for passing his school exams, when he was 11. He purchased his first bottle of poison, 25g of sodium antimony tartrate, from a local pharmacist when aged 13½, even though sales of scheduled poisons were prohibited to those under 17. The pharmacist was duped into thinking Young was much older owing to his extensive chemical knowledge. Before this point Young had only studied the underlying theories of chemistry, and particularly of poisons, but in early 1961 he started putting his theories into practice. His victims were his father Fred, his sister Winifred, a school friend, Chris Williams, and Molly, his stepmother, who was singled out for particular attention because he disliked her.

Young chose his victims because it was easy to poison them, rather than because he had any particular grudge. He tampered with food in the family house, dropping antimony compounds into tea and coffee or sometimes into jars of sauce or chutney. The effects were dramatic. Antimony compounds have been used for centuries as emetics; their ingestion results in copious and violent vomiting. In this respect antimony is its own antidote; much of the poison is expelled shortly after ingestion. Unfortunately the small amount that remains in the body stays for a long time, and repeated doses of antimony salts have a cumulative effect. A lethal dose for an adult is around 1g.

Young's antimony-based poisonings continued over several months, but despite the victims consulting numerous doctors and specialists, poisons, and indeed Graham, were never suspected. One morning Young's sister Winifred was drinking her usual cup of tea, but she complained that it tasted bitter. She didn't finish the tea, and went off to work as normal. On the way she started to feel dizzy, and she had to be helped off the bus. Somehow she got to work, where she found she could not focus her eyes properly. Her co-workers were worried, and one of them took her to a nearby hospital where she was diagnosed, and treated for atropine poisoning. This time Young was immediately suspected, and there was a huge row when Winifred arrived home. Graham flatly denied poisoning his sister, and became so upset that Winifred eventually apologised to him. Later he confessed to adding 50mg of atropine to her tea (the lethal dose is approximately 100mg).

Meanwhile Molly's symptoms were getting worse, and she was hospitalised on more than one occasion. The repeated doses of antimony compounds were not having their desired effect, though, and Young suspected that Molly was developing a tolerance to the poison. He decided to change tactics, and on 20 April 1962 he added 1,300mg of a thallium salt to her evening meal. The following day she woke up with a stiff neck, and pins-and-needles in her hands and feet. Her symptoms got worse over the course of the day, and when Fred came back from a lunchtime pint he found Molly in the garden, writhing in agony, as Graham watched her from the kitchen window. Molly was rushed to hospital, but she died later the same day. Molly's death was attributed to natural causes, and an inquest was not thought necessary. Graham Young helpfully suggested that her remains should be cremated.

Thinking that he had got away with it, Graham now turned his attention to his father Fred. He would accompany his father to the pub on Sunday evenings, and slip a dose of antimony into his pint when he went off to the toilet. Fred soon became very ill, and he was eventually hospitalised, where he was diagnosed with either arsenic or antimony poisoning. By now 15 years old, Graham offered advice to the doctors on how they could distinguish between the two poisons. The family was now genuinely suspicious of Graham, and his father banned him from his hospital bedside.

Back at school, Young was attracting the attention of his teachers through his incredible knowledge of chemistry, the only subject he excelled at despite being a bright student. His fascination with poisons worried the staff, and the long-term illness of Graham's school friend Chris Williams began to look suspicious. A psychiatrist was called in, who questioned Young on the pretext of a careers interview. She grew convinced that he was a psychopathic poisoner, and the following day the police were called. Graham Young's arrest led to the discovery of a stash of poisons and poison books in his bedroom. On questioning, he also confessed to having more secret stores of chemicals hidden not far from the family home.

Graham Young was charged with poisoning Chris Williams, and Fred and Winifred Young (no mention was made of Molly's death). He was sent to Broadmoor high-security psychiatric hospital, with no question of his release for 15 years without the express permission of the Home Secretary – at the time, he was the third-youngest admission in the hospital's history. During his time at Broadmoor Young appeared to reform and became a model patient, in spite of several poisoning incidents that occurred during his stay. The first, the suicide of John Berridge using cyanide, occurred within a month of Young's arrival. Although Young was never conclusively proven to have been responsible, he was believed by fellow inmates to have distilled cyanide from the leaves of the laurel bushes that grew abundantly around the perimeter of the hospital grounds. It couldn't be proven that this was the source of the cyanide, but the laurel bushes were cut down all the same. Young is also thought to have added toilet detergent to the nurses' coffee, and sugar soap (a powerfully alkaline cleaning product) to the tea urn. Luckily both of these additions were discovered before anyone took a drink.

Young's apparently reformed behaviour was a show, put on so he could obtain an early release. The chief psychiatrist was fooled and declared him to be no longer a threat. But before he was released, Young told a nurse 'I'm going to kill one person for every year I've spent in this place.' At the time he had been in Broadmoor for eight years; it was less than a month after his release before he began to put his plan into action.

Young got a job at Hadland's, a company that produced cameras and photographic equipment. At his interview for the position he explained his eight-year absence from society as being due to a mental breakdown, from which he was now fully recovered. The interviewers sought (and received) confirmation from Young's psychiatrist that he was completely well; no one told Hadland's the real reason for this gap on his curriculum vitae, or where he had been.

Coincidentally, thallium was a key ingredient in the glass of the camera lenses produced at Hadland's, though no thallium compounds were kept on site – Young had to travel to London to obtain his stocks. At work Young took on the responsibility of collecting his colleagues' tea from the tea trolley in the corridor. Each employee had their own mug, and for a brief part of his journey back from the tea trolley Young was unobserved. He always carried poison with him.

Young did as he had promised and poisoned eight people after leaving Broadmoor; two of his victims, Bob Egle and Fred Biggs, died. These eight people were examined by a combined total of 43 physicians during their illnesses, from GPs to specialists, and not one of them diagnosed poisoning. Young used both antimony and thallium compounds during his time at Hadland's. The symptoms displayed by his victims included vomiting, stomach pains, pains in the feet developing to pains all over the body so great that even the weight of bedclothes became unbearable, paralysis to the extent that victims could not speak, terrifying hallucinations and mental confusion. Fred Biggs started to go blind.

Bob Egle survived eight days of agony after he drank his poisoned tea before he died. His death was attributed to broncho-pneumonia complicated by Guillan-Barre syndrome, a disease of the nerves so rare that one of his kidneys was preserved for further study. Bob Egle's body was cremated, but later examination of his ashes and his preserved kidney showed traces of enough thallium to cause his death. Fred Biggs lasted 20 days before finally succumbing to the poison in his body. Biggs's post-mortem was conducted in the full knowledge that he had been poisoned with thallium, but initially no trace could be found in his remains. Days later, after further tests on tissue from Biggs's body, thallium was finally identified.

The spate of illnesses at the factory had been nicknamed the 'Bovingdon Bug' after the village where the factory was located. After Fred Biggs's death, managers at the factory became so concerned that they called in a team of doctors to investigate; a viral infection of unknown origin seemed to them the likeliest explanation (though they considered heavy-metal poisoning from nearby contaminated land as an improbable alternative). At a meeting called for all the employees to discuss the 'bug', Young stood up and talked at length about thallium and its symptoms. He suggested thallium was far more consistent with the symptoms displayed by the victims than the doctors' theory of a virus affecting the nerves. His behaviour was suspicious, to say the least; Young was also one of the few people working in the store room who was completely unaffected by the 'bug'.

The doctor in charge of the investigation questioned Young, and found his knowledge of medicine extended only to toxicology. Further investigations into Young's past finally revealed his time at Broadmoor. He was arrested, and during his time in police custody boasted that he had got away with the perfect murder in 1962 when he had poisoned his stepmother. The following day he made a full confession, and even criticised the treatment Jethro Batt, another poison victim, was receiving in hospital, suggesting that dimercaprol and potassium chloride should be given as an antidote.

A search of Young's room found bottles of poison in drawers and along the window ledge. He had an extensive library devoted to poisons, poisoners and toxicology. Under his bed

was the most telling piece of evidence, a diary, where Young
had kept an account of who he poisoned, which poison he
used, and when, as well as the results of his work. In the dock
Young claimed that the diary represented notes for a novel he
was writing. The jury took an hour to find Young guilty, with
most of this time spent sorting out the many charges against
him. He was sentenced to life imprisonment, and died in
prison in 1990.

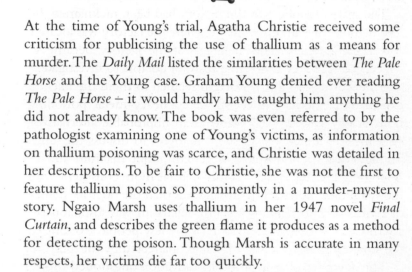

At the time of Young's trial, Agatha Christie received some
criticism for publicising the use of thallium as a means for
murder. The *Daily Mail* listed the similarities between *The Pale
Horse* and the Young case. Graham Young denied ever reading
The Pale Horse – it would hardly have taught him anything he
did not already know. The book was even referred to by the
pathologist examining one of Young's victims, as information
on thallium poisoning was scarce, and Christie was detailed in
her descriptions. To be fair to Christie, she was not the first to
feature thallium poison so prominently in a murder-mystery
story. Ngaio Marsh uses thallium in her 1947 novel *Final
Curtain*, and describes the green flame it produces as a method
for detecting the poison. Though Marsh is accurate in many
respects, her victims die far too quickly.

The *Daily Mail* may have been critical of Agatha Christie,
but her accuracy was vindicated a few years later. In 1975
Christie received a letter from a woman living in South
America. The letter-writer had realised that she was witnessing
the slow poisoning of a man by his young wife, and recognised
the symptoms of thallium poisoning because she had read *The
Pale Horse*. The letter concluded, 'But of this I am quite, quite
certain – had I not read *The Pale Horse* and thus learned of the
effects of thallium poisoning, X would not have survived; it
was only the prompt medication which saved him; and the
doctors, even if he had gone to hospital, would not have
known in time what the trouble was.'

Another case occurred in 1977, a year after Christie's death. A 19-month-old child was taken ill in Qatar. Her symptoms stumped all the doctors that examined her. The child's condition worsened and her parents took her, barely conscious, to London to be seen by specialists, but still the illness was mystifying. Then a nurse, Marsha Maitland, who had been reading *The Pale Horse,* suggested thallium poisoning. Urine samples were sent to Scotland Yard's forensic laboratory for analysis, and the presence of thallium was confirmed. Treatment was started and in two weeks the child's condition had stabilised; in three weeks she was noticeably improved, and four months later she was almost fully recovered. The source of the poison was an insecticide the parents had used to kill cockroaches and rats in the drains and cess-pit on their property in Qatar. Unknown to her parents, the child had found and consumed some of the insecticide. The physicians treating the young girl published a paper in the *British Journal of Hospital Medicine* discussing the symptoms, diagnosis and treatment of the case. At the end of the report, the authors acknowledged their indebtedness 'to the late Agatha Christie for her excellent and perceptive clinical descriptions, and to Nurse Maitland for keeping us up to date on the literature'.

Agatha and thallium

In *The Pale Horse,* Mrs Davis, who works for a customer survey company, notices that many of the people she had spoken to have died recently. Mrs Davis passes on a list of names to a priest in her final confession, before she too succumbs to a mysterious, fatal illness. Her list of names becomes the central clue in *The Pale Horse.* All of the people on the list had died recently, but the deaths were all thought to be due to natural causes. Attributed to pneumonia, brain tumours, toxic polyneuritis (damaged nerves), encephalitis (inflammation of the brain) and cerebral haemorrhage, the deaths seemed to bear no connection, and to have no single cause. The symptoms of two of the victims, Ginger Corrigan and Mrs Davis, are described in a little detail. Mrs Davis's symptoms began with 'flu', so she takes a few days off work, and during that time she appeared to make a recovery. Still not completely well, Mrs Davis returns to work, but two days later she can barely climb the stairs, has difficulty breathing and has a high fever. She dies the same day.

In an effort to trap the murderer, or at least discover their method, Mark Easterbrook poses as a client of the Pale Horse organisation, and his friend Ginger Corrigan volunteers to pose as an inconvenient ex-wife that Easterbrook wishes out of the way. To ensure nothing happens to Ginger, she hides herself away in a flat in London to watch events unfold. Meanwhile, Easterbrook attends a ceremony at the Pale Horse Inn in Much Deeping, miles from London. There are a few theatrical tricks but nothing to suggest that Ginger is in any danger. The 'witches' at the ceremony do not even ask the name of the intended victim, and never leave the village – how could they possibly hurt Ginger? Yet somehow Ginger is affected. Her symptoms start with 'a bit of a sore throat … I'm starting a cold I expect, or a touch of 'flu … aching all over'. A doctor is called, who diagnoses a touch of flu and sends her to bed. Later, the aching gets worse – 'everything hurts'. By now Ginger cannot bear anything touching her. When her condition deteriorates further she is taken to a nursing home and diagnosed with broncho-pneumonia following influenza, but with some unusual symptoms. Later still, Ginger's hair starts to fall out.

A chance observation reveals the clue to solving the case. Easterbrook sees the vicar's wife treating her dog for ringworm, rubbing some cream on the affected area to make the fur fall out; hair falling out is the only common symptom in all the deaths associated with the Pale Horse organisation. By lucky chance, Easterbrook had read an article about a case of industrial thallium poisoning in the United States. A lot of workers in a factory died one after another, but their deaths were attributed to a range of natural causes – paratyphoid (similar to typhoid but caused by different bacteria), apoplexy, alcoholic neuritis, bulbar paralysis (paralysis of tongue and muscles controlling the vocal cords), epilepsy and gastro-enteritis. The industrial case described in the novel was invented but it is similar to incidents in the Netherlands in the 1930s. Christie, through Easterbrook, also discusses a thallium-based murder case (again in the United States) to highlight the wide range of symptoms thallium poisoning can present. The case was probably invented, but there was a spate of thallium poisonings in Australia in the 1950s that Christie could have drawn upon.

Mark Easterbrook alerts doctors and the police, and thallium poisoning is confirmed. There was no standard test for thallium in a murder investigation in 1961 as there had never been a recorded case of thallium being used for this purpose in Britain. Thallium was so poorly understood at the time that the causes of death would not have appeared suspicious, and even if a pattern had been detected, almost certainly the authorities would not have suspected thallium of being the cause. While the victim is still alive urine can be analysed to detect thallium, using flame spectroscopy. Thallium levels would have to be above a certain level to confirm a poisoning case, though, as it is such a common element that everyone has trace amounts in their bodies. Luckily these traces of thallium never naturally build up to dangerous levels.

Pointing the finger at thallium grows a little more difficult at a post-mortem. Aside from the loss of hair, there are few common symptoms that can be relied on. Internally there are no characteristic signs in the body to show the presence of this metal, and the damage caused by thallium can easily be attributed to natural diseases. Analysis of body tissue to prove the presence of thallium can also be very difficult. Thallium distributes itself so well throughout the body that only tiny amounts may be found in any particular organ. As analytical techniques have improved, the detection of smaller and smaller amounts of thallium has become possible.[*]

_____

[*] As we saw earlier, thallium has even been identified in the cremated ashes of murder victims, such as Bob Egle in the Young case.

The method used to murder so many people in *The Pale Horse* is frighteningly easy, and it is a surprise that there had been no real-life cases in Britain before the book's publication. In the book, the Pale Horse organisation is an agency for contract killings. Someone with designs on advancing an inheritance or disposing of a despicable relative would contact an agent, who would take the details of the victim. A trip to the Pale Horse Inn in Much Deeping would see a sham seance, designed to give the impression of supernatural forces at work. Meanwhile the victim would be asked to take part in a customer survey on the types and brands of household products they used; later, they would receive another visitor posing as a gas-man, or an inspector from the electricity company. The visitor substitutes one household product for another, identical in appearance, but laced with thallium salts; it could be a type of tea, a cosmetic, a medicine, a soap or even a shampoo. The victim wouldn't even have to receive a fatal dose in one go, as the cumulative effects of regular doses of thallium would eventually lead to the same end result: a slow and agonising death.

Lord Edgware Dies

To die, to sleep,
To sleep, perchance to Dream; ay, there's the rub;
For in that sleep of death, what dreams may come,
When we have shuffled off this mortal coil,
Must give us pause.

William Shakespeare, *Hamlet*

The use of barbiturates as poisons dates Agatha Christie's writing as much as the clothes and cars described in her work. Barbiturates enjoyed a brief and notorious period in the limelight, between around 1920 and 1970. Initially they were considered a safe and effective sedative, but by the 1960s their dangers had been recognised and barbiturate drugs were slowly phased out of regular use as safer alternatives became available. Veronal was the trade name for the first barbiturate drug, released onto the market in the early 1900s. As the doctor in

Lord Edgware Dies points out, 'Veronal's uncertain stuff. You can take a devil of a lot and it won't kill you, and you can take very little and off you go. It's a dangerous drug for that reason.'

Christie described the use of barbiturates in several of her stories but employed them only four times for murder; two of her characters used barbiturates to commit suicide, an appropriate choice given that barbiturate overdose was a common way of taking one's own life at the time she was writing.

The 1933 Christie novel *Lord Edgware Dies* gives a most detailed description of the use of Veronal. The story centres around the death of the title character, who is found stabbed. The last person to visit Lord Edgware was his estranged wife, Jane Wilkinson, an American actress and rising star. The next day the papers report the discovery of Lord Edgware's body. They also say that Jane Wilkinson had attended a prominent dinner party the previous evening, giving her a seemingly unbreakable alibi.

The American title for the book, *Thirteen at Dinner*, refers to the superstition that if there are 13 people sitting at a dinner table, the first to rise will die. The dinner party where Wilkinson was seen had 13 guests, and she was apparently the first to rise. Luck seems to have been on her side, though, as it is another actress, Carlotta Adams, a young American actress and brilliant impersonator, who is subsequently found dead. Carlotta's character was based on the American actress Ruth Draper, famous for her character-driven monologues, but the character's death has many similarities to that of another real-life actress, Billie Carleton.

The story of Billie Carleton has some parallels with *Lord Edgware Dies*, and it certainly inspired the Agatha Christie short story *The Affair of the Victory Ball*. Billie Carleton was a young actress and singer who, in 1918, died in what was perhaps the first showbiz sex and drugs scandal. She attended a 'Victory Ball' at the Royal Albert Hall in London, one night in

November 1918. The party continued at Carleton's flat until the morning, but by 10 a.m. her guests had finally left, and she made a telephone call, the last conversation she ever had. At 11.30 a.m. Carleton's maid arrived to find her snoring. By 3.30 p.m. the snoring had stopped. A doctor was called, and he attempted to resuscitate her with an injection of brandy and strychnine. The attempt failed. Though cocaine overdose was the accepted verdict at the time, it has since been suggested that Carleton in fact died of the barbiturates she had been prescribed to cope with her cocaine 'hangovers'.

In *Lord Edgware Dies,* Carlotta's death is attributed to an overdose of the barbiturate Veronal. It is initially thought to have been an accident because a jewelled box of Veronal powder found in Carlotta's bag suggests that she was a regular user and, on this occasion, had taken too much. With one murder, one accidental death and an actress who appears to have been in two places at the same time, it's a puzzle that requires the genius of a legendary Belgian detective to solve it.

The story of barbiturates

Barbiturates are derivatives of barbituric acid, a compound first prepared by Adolf von Baeyer (1794–1885) in 1864. There are two stories that supposedly explain the naming of the acid. One has it that Baeyer was obsessed with a woman named Barbara, and he named it after her; the other holds that Baeyer made his discovery on St Barbara's Day. By 1900, around 2,000 variations on the basic barbituric acid unit had been created and tested; some of these were found to have beneficial effects on humans. These compounds form when barbituric acid is 'cyclised', to form a ring structure made of two nitrogen atoms and four carbons. This basic structure can be modified by changing the groups attached to one of the four carbon atoms or the two nitrogen atoms. A bewildering number of variations is possible.

From the creation of the first barbiturates in the early 1900s, more and more were synthesised and released onto the market. They soon became the standard treatment for depression and insomnia.

Changes in structure can change the solubility of a compound in water and fats, and thereby change how readily the drug is absorbed into the bloodstream, as well as how easily it can cross the blood-brain barrier. A change in structure will also affect the strength of its binding to target sites in the nervous system, meaning effects can take longer or shorter times to kick in. The rapidity with which these groups can be removed by the body and rendered inactive also leads to different periods of sedation. Until barbiturates were produced, the only sedative drugs available had been bromides (see page 250). Bromides had a very unpleasant taste, and were only moderately effective. They also led to many more side effects than barbiturates.

Barbiturates are white powders that would have been sold as tablets or as loose powder to be dissolved in water. They had a slightly bitter taste, but the tablets could be swallowed quickly (and barbiturates could be injected, though this was rare). Barbiturates were sold under a variety of trade names, but the actual names of the drugs were more formulaic – in the United States the name usually ended in 'al', while everywhere else barbiturates could be identified by a name ending in 'one'. For example, barbital and barbitone are the same compound – and this was marketed in the United States under the name Veronal. Veronal was widely prescribed as a sleeping drug because in low doses it acted as a sedative, and at higher doses as a hypnotic to render the taker into a state something like sleep.[*]

The dampening effect barbiturate drugs have on the nervous system means they also have applications in other areas. In 1911 a German doctor, Alfred Hauptmann (1881–1948), was living and working in a hospital that had a ward of epileptic patients. The noises made when the patients were suffering seizures kept Hauptmann awake at night. In a desperate attempt to get some rest he decided to sedate the epileptics with phenobarbitone. The drug succeeded in sedating them, and it

[*]Veronal was mostly sold in the United States, but in Britain the sodium salt of barbitone was available, under the name Medinal.

also reduced their fits, with this benefit continuing even after the sedative effects had worn off. Phenobarbitone is still used as an anticonvulsant drug in the treatment of epilepsy today.

Other barbiturates included Pentothal, which found notoriety as a 'truth drug'.* It is thought that lying is a more complex procedure than truth-telling; so, in theory, by using a drug that depresses the function of nerves in the cortex of the brain, people are more likely to tell the truth. Pentothal is also a good example of the potentially lethal nature of barbiturates; it has been used in the United States in lethal injections of criminals, as have other barbiturate drugs, the idea being that large doses of these compounds induce a deep coma, allowing the administration of other lethal compounds that stop the heart.

When barbiturates were first released onto the market there was thought to be a safe margin between a therapeutic dose and a lethal one. However, the 12-fold increase in suicides by barbiturates between 1938 and 1954 told a different story. Veronal was prescribed in therapeutic doses of between five and fifteen grains (0.3–1.0g), but a lethal dose was only around sixty grains (around 4g). In the 1912 edition of *The Art of Dispensing* that Dame Agatha studied for her dispensing exams, a dose of seven and a half grains (about 0.5g) of Veronal for men and five grains (about 0.3g) for women is recommended. The gap between therapeutic and lethal doses could be considerably narrowed because, over prolonged use, individuals can develop a tolerance to the drug, and they go on to require larger doses to achieve the same sedative effect. These drugs are also potentially addictive in a similar way to alcohol, opening them up to abuse. Also, owing to the fact that barbiturates slow reactions and increase drowsiness, many sleepy individuals took an extra (lethal) dose without realising what they were doing.

*Pentothal (aka sodium thiopental) may make the subject more cooperative and talkative, but the value and reliability of what they may divulge is questionable.

In 1948, worldwide production of barbiturates was more than 300 tons per annum, but in the 1950s the dangers had begun to be better understood, as an increasing number of people became addicted and some died as a result of overdose. The next decade saw barbiturates largely replaced by benzodiazepines, which produce much the same effects (and may also be addictive) but are far less likely to lead to accidental overdose. One of the few remaining medical applications for barbiturates is in surgery; rapid-acting barbiturates that yield short periods of sedation are administered before an operation to induce unconsciousness.

How barbiturates kill

All barbiturates exhibit a similar spectrum of pharmacological effects, but they differ with respect to the time they either kick in or last for. Their effects are due to interactions with the nervous system that make it more difficult for nerve cells to be activated, resulting in an overall depression of nerve activity.

Barbiturates activate specific sites called gamma-aminobutyric acid (or GABA) receptors in the nerve cells. GABA is one of the many neurotransmitters, or molecules used as chemical messengers between nerves. GABA has other functions within the body, including maintaining muscle tone, but its principal role is to inhibit nerve activation.

Barbiturates mainly interact with GABA receptors in nerve cells in the brain. Each GABA receptor is formed of five subunits around a central pore that allows chloride ions (Cl^-) to flow into the nerve cell. There are 15 different potential subunits; each of the five making up a receptor can be different. Therefore, there is huge variety in the composition of GABA receptors; this helps enable the diversity and complexity necessary for the operation of an intricate and sophisticated organ such as the brain.

As we have seen, signals are generated by the movement of potassium ions in and sodium ions out of nerve cells (see page 145). The movement of these ions produces a tiny electric current that is transmitted along the length of the nerve, triggering the release of neurotransmitters at its end to pass on the signal to an adjoining nerve or muscle cell. When at rest, nerve cells are slightly negatively charged on the inside. When the nerve 'fires', the movement of ions means the internal charge goes from negative to zero, and then – for just a few milliseconds – becomes slightly positive. Molecular pumps then move the ions back to their original positions, restoring the negative charge inside the nerve cell and making it ready to fire again. Barbiturate interactions with the cell encourage more negative charge to accumulate inside when it is at rest, making it more difficult for the nerve to fire. In this way, nervous function is depressed.

GABA receptors also interact with alcohol; when taken with barbiturates, the two enhance each other's effects, leading to a greater sedation than the sum of the parts might suggest. Benzodiazepines also bind to GABA receptors. The binding of one of these molecules increases the binding of another and taking combinations of these drugs can therefore be very dangerous.

By suppressing the activity of the central nervous system, barbiturates can cause drowsiness and slowed reactions. This has been described as like being drunk, especially if there is a delay after taking the drug before going to sleep. The end result is an unconscious state that resembles sleep. There are three distinct stages of sleep; humans normally go through all of them over the course of the night, spending around 20 per cent of their sleeping time in 'deep' sleep, 20 per cent in 'dream' sleep and 60 per cent in 'light' sleep. Barbiturates change the sleep pattern, resulting in less dream and deep sleep but more light sleep (approximately 80 per cent). People taking sleeping drugs often report that they do not feel as rested after drug-induced sleep as they do after normal sleep. They may find it difficult to wake up, and remain groggy the following morning. In the case of barbiturates this is sometimes described as a 'hangover'.

When people stop taking sleeping drugs their sleep is often disrupted further as the sleep pattern changes again. After the drug is withdrawn they experience a large increase in the time spent in dream sleep (to 40 per cent) and a decrease in deep sleep (to 10 per cent). Nightmares and vivid dreams are a common consequence. After a sudden withdrawal from barbiturates, especially following prolonged use, the symptoms are more serious and can even kill. Individuals can experience extreme anxiety, convulsions and hallucinations, particularly if they have been a heavy user of the drug. Other symptoms may include nausea and vomiting, and in extreme cases delirium, fever and coma.

When an overdose of barbiturates is taken, the activity of the central nervous system is depressed to the point that breathing may stop. However, there are other effects that can contribute to or cause death. Barbiturates cause fluids to build up in the muscles, lungs and brain, causing oedema and pneumonia. Slower breathing rates lead to higher levels of carbon dioxide in the body, as this cannot be expelled so effectively from the lungs. Carbon dioxide in the blood forms carbonic acid, increasing its acidity. There may also be a lack of oxygen being taken into the body, and cyanosis – a blue coloration of the skin – may be seen. Due to suppression of the 'cough' reflex, it will be difficult to clear fluid from the lungs and throat; this can be especially dangerous if the barbiturates cause the individual to vomit.

Barbiturates also interact with enzymes in the liver that are involved in metabolising drugs, including the barbiturates themselves. Prolonged use of barbiturates results in the faster metabolism of the drug into inactive forms (which are excreted from the body). The result of this is the development of tolerance. Ever-increasing doses of barbiturates are needed to keep pace with the breakdown in the liver while still having enough left to act on the nerves. This also means that barbiturates are dangerous when taken with other medication, as the increased rate of metabolism results in other drugs having less effect on the body. Barbiturates are dangerous drugs; the combined effects of a build-up of tolerance and their sedative effects make a patient mentally sluggish, at which point they are more likely to accidentally overdose. This appears to be the case with Carlotta Adams in *Lord Edgware Dies*.

Is there an antidote?

There is no specific antidote for barbiturate overdose, but with supportive care more than 95 per cent of patients would be expected to make a full recovery. In severe overdoses it may take up to five days for the patient to regain consciousness, but by supporting their breathing, ensuring that they are receiving enough oxygen and that carbon dioxide is being expelled from the lungs, as well as clearing their lungs of mucus, they are unlikely to die.

Some real-life cases

There have been many famous victims of barbiturate overdose: Judy Garland and Jimi Hendrix, for example. These deaths have generally been attributed to accident or suicide. One of the most famous victims of a barbiturate overdose was Marilyn Monroe. At Monroe's autopsy the pathologist found enough Nembutal (pentobarbital) and chloral hydrate in her body to kill ten people, but debate still rages as to how it got there.

The first known case of murder using barbiturates occurred in 1955, long after Agatha Christie wrote *Lord Edgware Dies*. She can hardly be blamed for inspiring this particularly cold-hearted murder, as the lethal properties of barbiturates were well known by the 1950s, owing to the number of suicide victims using these drugs.

At 1.30 p.m. on 21 July 1955, John Armstrong called Dr Bernard Johnson, saying his five-month-old baby son Terence was ill. Dr Johnson's colleague, Dr Buchanan, had received a similar call the previous evening, but Armstrong had not seemed overly worried, so Buchanan waited until 9 a.m. the following morning before calling round to find the baby happy and well. Armstrong had two other children with his wife Janet and the couple, residents of Gosport in Hampshire, were feeling the financial strain of supporting the family. They had also suffered a tragedy when their first son, Stephen, had died the previous year. Their daughter Pamela had also suffered an illness serious enough to need a stay in hospital. The cause of the illness was not known at the time, but Pamela made a full recovery and returned home.

A second call to the doctor on 22 July sounded more urgent, and Dr Johnson went directly to the house, where he found the baby dead. Initially he did not suspect foul play but the parents' apparent lack of grief was troubling,* and in any case he could not determine the cause of death. A post-mortem was ordered, as would be the case for any sudden death of a child today. Dr Johnson also took the precaution of taking possession of the baby's bottle, and a pillow that Terence had vomited on to the previous night.

The post-mortem examination revealed no obvious cause of death, but a shrivelled red shell was found in the larynx of the baby, and more red shells lay among the stomach contents. The pathologist thought the shells looked like the skins of daphne berries, which are highly toxic, and a fruiting daphne tree was growing in the Armstrongs' garden. Terence had been in his pram under the tree the day before his death.

The pathologist carefully placed the red shells from the larynx in a bottle of formaldehyde and stored it in a refrigerator, along with another bottle containing the stomach contents of the child, for further examination at a later date. The next day the pathologist took a look at the bottles and found that the red shells had disappeared, leaving the liquid contents stained red.

Chemical analysis of the liquids revealed the presence of corn starch and eosin, a red dye, but no signs of daphne berries or their toxic components, or any other poison, for that matter. The cause of the infant's death remained a mystery, and it would have stayed that way if it hadn't been for the bad impression Armstrong had made on the investigating officer.

---

*When the coroner's assistants went to the house to look for a daphne tree they found the parents calmly watching television, as if nothing had happened.

The detective made further enquiries in Armstrong's workplace, and went back to see the pathologist.

In the intervening period the pathologist had mulled over the question of the disappearing shells. The colour of the shells reminded him of the gelatin capsules used to contain the drug Seconal.* Seconal was freely prescribed in the 1950s. It was first released on to the market in 1934, and by the 1960s it was widely abused; the red capsules were commonly known as 'seccies', 'red devils' or 'red hearts'. Seconal is still manufactured today; it is prescribed in 100mg tablets for the treatment of epilepsy, as a temporary treatment for insomnia, and pre-operatively in short surgical procedures. But John Armstrong had no need of a prescription, as he was a nurse at the local hospital and could help himself. Enquiries at the hospital revealed that several boxes of Seconal had disappeared recently.

The pathologist obtained some capsules and showed that they dissolved in gastric juices, producing the same red colour as in the dissolved shells from the baby's body; Seconal, though never previously used in a murder, would be expected to kill a baby in very small quantities. By this stage the scientific expertise of Scotland Yard had been called in. Seconal, once extracted from the vomit on the baby's pillow, could be identified by its unique melting point (among the barbiturates) of 95°C.

With circumstantial evidence against John Armstrong mounting, an investigation was opened into the death of his eldest son Stephen and Pamela's illness the previous year. The symptoms all the children had displayed were remarkably similar, with difficulty breathing, discoloration to the face and drowsiness. Stephen's body was exhumed but it was found to be too decomposed to determine whether Seconal was present or not; however, the police and pathologists were convinced that both he and baby Terence had been killed with barbiturates. The only thing that could not be proved was that the

*Seconal was the barbiturate implicated in the death of Judy Garland.

Armstrongs had been in possession of Seconal on the day the baby died. Evidence of this came a year later, and in a rather unusual manner.

In 1956 Janet Armstrong applied for a divorce from her husband; she accused him of beating her regularly. When the court refused to grant the divorce she went to the police and offered a statement. Janet asserted that her husband had told her to get rid of all the red capsules in the house three days after Terence's death, which she had done. Only later, when the cause of the baby's death had been revealed, did she become suspicious of her husband. She had not told the police at the time because she was terrified of another beating. John Armstrong was subsequently found guilty,[*] but the trial raised concerns about the detection of barbiturate drugs in murder cases. The death of the Armstrongs' baby would probably have been dismissed as due to unknown natural causes if it had not been for the bad impression the baby's parents had given to those investigating the murder.[†]

Agatha and Veronal

In *Lord Edgware Dies*, Carlotta Adams's death is carefully staged to look like an accidental overdose. The actress drinks a Veronal-laced toast with her murderer a few hours before she dies; a bitter-tasting drink would have disguised the slightly bitter flavour of the barbiturate. Veronal's symptoms start to kick in up to an hour after ingestion, meaning that Carlotta had time to arrive back at her flat before beginning to feel the effects. She tries to make a phone call, but the Veronal makes

---

[*]Armstrong was sentenced to death, but this was later reduced to life imprisonment. After the trial, Janet confessed to giving the child a capsule as she believed it would help make him sleep. The Home Secretary at the time considered reopening the case, but Janet couldn't be tried again. In the end it was decided that a single capsule would not have resulted in Terence's death. John Armstrong was still guilty of murder.

[†]Motives for these crimes never seem to have been established.

her tired so instead she goes to bed and drinks a glass of hot milk, prepared by her maid, Alice Bennett, before going to sleep. Carlotta is unlikely to have taken the drug herself when she got home; if she had, the feelings of drowsiness would have presented themselves after she had attempted to make the phone call. If she was already sleepy when she tried to make the call why would she have taken a dose of Veronal? For similar reasons we can rule out the glass of milk as the vehicle for the fatal dose. Also, Alice drinks the same milk the following morning with no ill effects.

Carlotta dies in her sleep during the night. Alice the maid finds her cold to the touch the next morning, so death must have occurred several hours before she is discovered. When the doctor arrives he quickly reaches the conclusion that her death was due to a Veronal overdose, because of the appearance of the body and the presence of the Veronal powder in a jewelled box in Carlotta's bag. There are no marks of a hypodermic syringe on the body, so the doctor concludes that Carlotta was not a drug addict. But the box does suggest that she had been a regular user. Her death is therefore assumed to have been accidental. Carlotta came home late at night feeling tired and 'strung up' by her recent performances so she took a dose of Veronal to help her sleep, and mistakenly took too much. However, neither Alice the maid nor the victim's sister Lucie Adams believe Carlotta took sleeping drugs. 'She had a horror of that kind of thing,' as Lucie remarks.

The dose of barbiturate she had been given was clearly sufficient to kill, but this need not have been a very large dose. If Carlotta's maid and sister were to be believed the actress did not use barbiturates, and therefore could not have developed a tolerance to the drug. By adding Veronal to an alcoholic drink the murderer was increasing its potency; just a few grams of the drug would have been enough to ensure that the sleepy Carlotta never woke up.

There is no mention of a post-mortem examination, but the doctor and police seem happy to accept the accidental overdose theory. Had her sudden death occurred without evidence of

Veronal (or a similar drug) in Carlotta's possession then a post-mortem might have been ordered. There are no characteristic signs of barbiturate poisoning on the body itself post-mortem – there may be indications of oedema, pneumonia and cerebral oedema, but these could be attributed to natural diseases – but toxicological analysis of the liver and stomach contents as well as the blood would have revealed the presence of barbiturates, even using 1930s techniques. This would have confirmed the presence of the drug. Vomit would be particularly useful for analysis, as it would be expected to contain the highest concentration of the drug if it had been administered by mouth. In *Lord Edgware Dies* it seems that Carlotta's death was not preceded by vomiting; if it had been, Alice the maid would have noticed that something was wrong a lot sooner.

Identifying which of the many barbiturates was present would have been difficult; even the melting points of different barbiturates can vary by only a couple of degrees, requiring very precise and careful experimentation on the part of the toxicologist. Today, detection and identification of barbiturates is much easier thanks to the development of chromatographic techniques, while an improved version of the Stas method is effective in extracting barbiturates from human tissue; even if a pathologist has to resort to an analysis of maggots that have fed on a cadaver, it is still possible to trace the drug.

In *Lord Edgware Dies*, there is no uncertainty over how Carlotta Adams died, but Poirot has serious doubts over the suggestion that she committed suicide. And, of course, he is right – Carlotta was murdered, and to find out the culprit you will have to read the book.

Agatha Christie's selection of barbiturates as the poison is an ideal choice, especially as the tragic death of the real actress Billie Carleton would have been familiar to many of her readers. Christie had no need to rely on exotic or obscure drugs to do away with her victim in this book. Barbiturates were regularly prescribed at the time, and readers may well have had first-hand knowledge of such sedatives. The drug's details are supremely accurate, even down to the American name of the drug that killed her American victim. *Lord Edgware Dies* is perfect for its period – and contains all the key ingredients for a classic murder mystery from the Queen of Crime.

Appendix 1:
Christie's Causes of Death

This table lists all the Agatha Christie novels and short stories in order of publication, and the cause of death in each of them. Agatha Christie's plays are not included, nor are the titles published under the name Mary Westmacott. Not all Christie books that were published in the UK were published in the US (and vice versa), especially the short-story collections, and these sometimes vary in composition. An interesting note: *Three Blind Mice* has never been published in the UK as it provides the basis for Christie's play, *The Mousetrap*, which is still running; the book's publication would give away the identity of the murderer.

In the table below, please note:

★	Suicide	★★	Attempted murder
★★★	Medication withheld	★★★★	Invented drug

UK title	Methods of murder	US title
The Mysterious Affair at Styles	Strychnine	*The Mysterious Affair at Styles*
The Secret Adversary	Chloral hydrate Cyanide★ Morphine★★	*The Secret Adversary*
The Murder on the Links	Stabbed Morphine★★	*The Murder on the Links*
The Man in the Brown Suit	Electrocuted Strangled	*The Man in the Brown Suit*

UK title	Methods of murder	US title
Poirot Investigates		*Poirot Investigates*
The Adventure of the 'Western Star'		The Adventure of the 'Western Star'
The Tragedy at Marsdon Manor	Shot	The Tragedy at Marsdon Manor
The Adventure of the Cheap Flat		The Adventure of the Cheap Flat
The Mystery of Hunter's Lodge	Shot	The Mystery of Hunter's Lodge
The Million Dollar Bond Robbery		The Million Dollar Bond Robbery
The Adventure of the Egyptian Tomb	Blood poisoning Strychnine Shot★	The Adventure of the Egyptian Tomb
The Jewel Robbery at the Grand Metropolitan		The Jewel Robbery at the Grand Metropolitan
The Kidnapped Prime Minister		The Kidnapped Prime Minister
The Disappearance of Mr Davenheim		The Disappearance of Mr Davenheim
The Adventure of the Italian Nobleman	Hit on head	The Adventure of the Italian Nobleman
The Case of the Missing Will		The Case of the Missing Will
	Trinitrine★★★	The Chocolate Box
		The Veiled Lady
		The Lost Mine
The Secret of Chimneys	Shot	*The Secret of Chimneys*
The Murder of Roger Ackroyd	Arsenic Veronal★ Stabbed	*The Murder of Roger Ackroyd*

UK title	Methods of murder	US title
The Big Four	Cyanide Throat cut Yellow Jasmine Electrocuted Run over Stabbed	*The Big Four*
The Mystery of the Blue Train	Strangled	*The Mystery of the Blue Train*
The Seven Dials Mystery	Chloral hydrate Shot	*The Seven Dials Mystery*

Partners in Crime
    A Fairy in the Flat/
        A Pot of Tea
    The Affair of the Pink
        Pearl
    The Adventure of the
        Sinister Stranger
    Finessing the King/
        The Gentleman
        Dressed in Newspaper
    The Case of the
        Missing Lady
    Blindman's Buff
    The Man in the Mist
    The Crackler
    The Sunningdale
        Mystery
    The House of Lurking
        Death
    The Unbreakable Alibi
    The Clergyman's
        Daughter/The Red
        House
	
	Stabbed
	
	Electrocuted
	Hit on head
	
	Stabbed
	
	Arsenic★★
	Ricin
		Partners in Crime
		    A Fairy in the Flat/
		        A Pot of Tea
		    The Affair of the Pink
		        Pearl
		    The Adventure of the
		        Sinister Stranger
		    Finessing the King/
		        The Gentleman
		        Dressed in Newspaper
		    The Case of the
		        Missing Lady
		    Blindman's Buff
		    The Man in the Mist
		    The Crackler
		    The Sunningdale
		        Mystery
		    The House of Lurking
		        Death
		    The Unbreakable Alibi
		    The Clergyman's
		        Daughter/The Red
		        House

(*Continued*)

UK title	Methods of murder	US title
The Ambassador's Boots The Man Who Was No. 16		The Ambassador's Boots The Man Who Was No. 16
The Mysterious Mr Quin The Coming of Mr Quin The Shadow on the Glass At the 'Bells and Motley' The Sign in the Sky The Soul of the Croupier The Man from the Sea The Voice in the Dark The Face of Helen The Dead Harlequin The Bird with the Broken Wing The World's End Harlequin's Lane	Strychnine Shot Shot Drowned Drowned Poison gas★★ Shot Strangled	*The Mysterious Mr Quin* The Coming of Mr Quin The Shadow on the Glass At the 'Bells and Motley' The Sign in the Sky The Soul of the Croupier The Man from the Sea The Voice in the Dark The Face of Helen The Dead Harlequin The Bird with the Broken Wing The World's End Harlequin's Lane
The Murder at the Vicarage	Shot Sedative★★	*The Murder at the Vicarage*
The Sittaford Mystery	Hit on head	*The Murder at Hazelmoor*
Peril at End House	Shot Cocaine★	*Peril at End House*
The Thirteen Problems The Tuesday Night Club Ingots of Gold The Blood-Stained Pavement The Idol House of Astarte	Arsenic Hit on head Stabbed	*The Tuesday Club Murders* The Tuesday Night Club Ingots of Gold The Blood-Stained Pavement The Idol House of Astarte

(Continued)

UK title	Methods of murder	US title
Motive *v.* Opportunity		Motive *v.* Opportunity
The Thumb Mark of St. Peter	Atropine	The Thumb Mark of St. Peter
The Blue Geranium	Cyanide	The Blue Geranium
The Companion	Drowned	The Companion
The Four Suspects	Pushed down stairs	The Four Suspects
A Christmas Tragedy	Bludgeoned	A Christmas Tragedy
The Herb of Death	Digitalis	The Herb of Death
The Affair at the Bungalow		The Affair at the Bungalow
Death by Drowning	Drowned	Death by Drowning
Lord Edgware Dies	Stabbed Veronal	*Thirteen at Dinner*
The Hound of Death		
The Hound of Death	House collapse Lightning	
The Red Signal	Shot	
The Fourth Man	Strangled	
The Gypsy	Poisoned	
The Lamp	Starvation Natural causes	
Wireless	Heart attack	
The Witness for the Prosecution	Crowbar	
The Mystery of the Blue Jar		
The Strange Case of Sir Arthur Carmichael	Cyanide★★	
The Call of Wings	Hit by bus Hit by tube train	
The Last Seance	Supernatural	
SOS		
Murder on the Orient Express	Stabbed	*Murder in the Calais Coach*

UK title	Methods of murder	US title
The Listerdale Mystery		
The Listerdale Mystery		
Philomel Cottage	Heart attack	
The Girl in the Train		
Sing a Song of Sixpence	Hit on head	
The Manhood of Edward Robinson		
Accident	Cyanide Arsenic Fell off cliff	
Jane in Search of a Job		
A Fruitful Sunday		
Mr. Eastwood's Adventure		
The Golden Ball		
The Rajah's Emerald		
Swan Song	Stabbed	
Why Didn't They Ask Evans?	Pushed off cliff Morphine Shot	*The Boomerang Clue*
Parker Pyne Investigates		*Mr. Parker Pyne, Detective*
The Case of the Middle-aged Wife		The Case of the Middle-aged Wife
The Case of the Discontented Soldier		The Case of the Discontented Soldier
The Case of the Distressed Lady		The Case of the Distressed Lady
The Case of the Discontented Husband		The Case of the Discontented Husband
The Case of the City Clerk		The Case of the City Clerk
The Case of the Rich Woman		The Case of the Rich Woman
Have You Got Everything You Want?		Have You Got Everything You Want?

(*Continued*)

UK title	Methods of murder	US title
The Gate of Baghdad	Stabbed Cyanide★	The Gate of Baghdad
The House of Shiraz	Fell from balcony	The House of Shiraz
The Pearl of Price		The Pearl of Price
Death on the Nile	Strychnine	Death on the Nile
The Oracle at Delphi		The Oracle at Delphi
Three Act Tragedy	Nicotine	*Murder in Three Acts*
Death in the Clouds	Snake venom Cyanide	*Death in the Air*
The A.B.C. Murders	Hit on head Strangled Stabbed	*The A.B.C. Murders*
Murder in Mesopotamia	Hit on head Hydrochloric acid	*Murder in Mesopotamia*
Cards on the Table	Stabbed Anthrax Septicaemia Drowned Shot Veronal Hat paint	*Cards on the Table*
Dumb Witness	Phosphorus Chloral hydrate★	*Poirot Loses a Client*
Death on the Nile	Shot Stabbed	*Death on the Nile*
Murder in the Mews		*Dead Man's Mirror*
Murder in the Mews	Shot★	Murder in the Mews
The Incredible Theft		
Dead Man's Mirror	Shot	Dead Man's Mirror
Triangle at Rhodes	Strophanthin	Triangle at Rhodes
Appointment with Death	Digitoxin Shot★	*Appointment with Death*
Hercule Poirot's Christmas	Throat cut	*Murder for Christmas*

UK title	Methods of murder	US title
Murder is Easy	Hat paint Pushed out of window Pushed in canal Septicaemia Run over Arsenic Hit on head	*Easy to Kill*
And Then There Were None / Ten Little Niggers	Drowned Cyanide Hit on head Hanged★ Shot Run over Axe Chloral hydrate Amyl nitrite★★★ Poisoned Starvation Exposure Medical malpractice	*And Then There Were None / Ten Little Indians*
		The Regatta Mystery
		The Regatta Mystery
	Stabbed	The Mystery of the Baghdad Chest
	Strychnine	How Does Your Garden Grow?
		Problem at Pollensa Bay
	Cyanide	Yellow Iris
	Stabbed	Miss Marple Tells a Story
	Shot	The Dream
		In a Glass Darkly
	Heart attack	Problem at Sea
	Stabbed	
Sad Cypress	Morphine	*Sad Cypress*

UK title	Methods of murder	US title
One, Two, Buckle My Shoe	Shot Medinal Procaine and adrenaline	The Patriotic Murder / An Overdose of Death
Evil Under the Sun	Strangled Arsenic	Evil Under the Sun
N or M?	Shot	N or M?
The Body in the Library	Strangled Digitalin★★	The Body in the Library
Five Little Pigs	Coniine	Murder in Retrospect
The Moving Finger	Cyanide Hit on head	The Moving Finger
Towards Zero	Hit on head Heart attack Shot with arrow	Towards Zero
Death Comes as the End	Pushed off cliff Poisoned Shot with arrow	Death Comes as the End
Sparkling Cyanide	Cyanide Natural gas★★	Remembered Death
The Hollow	Shot Cyanide★	The Hollow / Murder After Hours

The Labours of Hercules
  The Nemean Lion
  The Lernaean Hydra
  The Arcadian Deer
  The Erymanthian Boar
  The Augean Stables
  The Stymphalean Birds
  The Cretan Bull
 	
  The Horses of
    Diomedes
  The Girdle of
    Hyppolita
	Strychnine★★
	Arsenic
	
	Stabbed
	
	Hit on head
	Atropine
	Shot★
		The Labours of Hercules
		  The Nemean Lion
		  The Lernaean Hydra
		  The Arcadian Deer
		  The Erymanthian Boar
		  The Augean Stables
		  The Stymphalean Birds
		  The Cretan Bull
		
		  The Horses of
		    Diomedes
		  The Girdle of
		    Hyppolita

(Continued)

UK title	Methods of murder	US title
The Flock of Geryon	Influenza/ typhoid/ gastric ulcer/ tuberculosis★★	The Flock of Geryon
The Apples of Hesperides	Fell	The Apples of Hesperides
The Capture of Cerberus		The Capture of Cerberus
Taken at the Flood	Hit on head Morphine Shot★	*There is a Tide …*
		The Witness for the Prosecution and Other Stories
	Arsenic	Accident
	Fell of cliff	
	Cyanide	
	Strangled	The Fourth Man
		The Mystery of the Blue Jar
		The Mystery of the Spanish Shawl (aka Mr. Eastwood's Adventure)
	Heart attack	Philomel Cottage
	Shot	The Red Signal
	Shot	The Second Gong
	Hit on head	Sing a Song of Sixpence SOS
	Heart attack	Where There's a Will (aka Wireless)
	Crowbar	The Witness for the Prosecution
Crooked House	Eserine Digitalis Car crash	*Crooked House*

UK title	Methods of murder	US title
		Three Blind Mice and Other Stories
	Neglect	Three Blind Mice
	Strangled	
		Strange Jest
	Strangled	The Tape-Measure Murder
		The Case of the Perfect Maid
	Strophanthin	The Case of the Caretaker
	Shot	The Third Floor Flat
		The Adventure of Johnny Waverly
	Pushed down stairs	Four-and-Twenty Blackbirds
	Hit on head	The Love Detectives
A Murder is Announced	Shot	*A Murder is Announced*
	Narcotics	
	Strangled	
They Came to Baghdad	Stabbed	*They Came to Baghdad*
		The Under Dog and Other Stories
	Hit on head	The Under Dog
	Stabbed	The Plymouth Express
	Stabbed	The Affair at the Victory Ball
	Cocaine	
		The Market Basing Mystery
	Formic acid	The Lemesurier Inheritance
	Arsenic	The Cornish Mystery
	Hit on head	The King of Clubs
		The Submarine Plans
		The Adventure of the Clapham Cook

UK title	Methods of murder	US title
Mrs McGinty's Dead/ Blood Will Tell	Hit on head Poisoned Strangled	*Mrs McGinty's Dead*
They Do It With Mirrors	Shot Aconitine★★ Crushed Drowned	*Murder with Mirrors*
After the Funeral/Murder at the Gallop	Hatchet Arsenic★★	*Funerals are Fatal*
A Pocket Full of Rye	Taxine Cyanide Strangled	*A Pocket Full of Rye*
Destination Unknown	Poisoned	*So Many Steps to Death*
Hickory Dickory Dock	Morphine Poisoned Hit on head Medinal	*Hickory Dickory Death*
Dead Man's Folly	Strangled Drowned	*Dead Man's Folly*
4.50 from Paddington	Strangled Arsenic Aconitine	*What Mrs McGillicuddy Saw*
Ordeal by Innocence	Hit on head Stabbed	*Ordeal by Innocence*
Cat Among the Pigeons	Shot Hit on head	*Cat Among the Pigeons*
The Adventure of the Christmas Pudding The Adventure of the Christmas Pudding The Mystery of the Spanish Chest The Under Dog Four and Twenty Blackbirds The Dream Greenshaw's Folly	Stabbed Hit on head Pushed down stairs Shot Shot with arrow	

UK title	Methods of murder	US title
	Cyanide★★	*Double Sin and Other Stories*
		Double Sin
		Wasps' Nest
		The Theft of the Royal Ruby (aka The Adventure of the Christmas Pudding)
		The Dressmaker's Doll
	Shot with arrow	Greenshaw's Folly
		The Double Clue
	Supernatural	The Last Seance
	Shot	Sanctuary
The Pale Horse	Thallium	*The Pale Horse*
	Hit on head	
The Mirror Crack'd from Side to Side	Calmo★★★★	*The Mirror Crack'd*
	Cyanide	
	Shot	
	Sleeping tablets	
The Clocks	Strangled	*The Clocks*
	Stabbed	
A Caribbean Mystery	Depressant drug	*A Caribbean Mystery*
	Stabbed	
	Atropine★★	
	Drowned	
At Bertram's Hotel	Shot	*At Bertram's Hotel*
	Car crash★	
Third Girl	Pushed out of window	*Third Girl*
	Stabbed	
Endless Night	Cyanide	*Endless Night*
	Drowned	
	Stabbed	
	Strangled	
By the Pricking of My Thumbs	Morphine	*By the Pricking of My Thumbs*

UK title	Methods of murder	US title
Hallowe'en Party	Drowned Stabbed Cyanide★	*Hallowe'en Party*
Passenger to Frankfurt	Shot Strychnine★★	*Passenger to Frankfurt*

	House collapsed
	Poisoned
	Starvation
	Natural causes
	Cyanide★★
	
	Hit by bus
	Hit by tube train
		The Golden Ball and Other Stories
		  The Listerdale Mystery
		  The Girl in The Train
		  The Manhood of
		    Edward Robinson
		  Jane in Search of a Job
		  A Fruitful Sunday
		  The Golden Ball
		  The Rajah's Emerald
		  Swan Song
		  The Hound of Death
		  The Gypsy
		  The Lamp
		
		  The Strange Case of Sir
		    Arthur Carmichael
		  The Call of Wings
		
		  Magnolia Blossom
		  Next to a Dog

Nemesis	Poisoned Crushed Strangled	*Nemesis*
Elephants Can Remember	Shot Hit on Head	*Elephants Can Remember*
Postern of Fate	Digitalis Hit on head	*Postern of Fate*

UK title	Methods of murder	US title
Poirot's Early Cases		*Hercule Poirot's Early Cases*
The Affair at the Victory Ball	Stabbed Cocaine	The Affair at the Victory Ball
The Adventure of the Clapham Cook		The Adventure of the Clapham Cook
The Cornish Mystery	Arsenic	The Cornish Mystery
The Adventure of Johnnie Waverly		The Adventure of Johnnie Waverly
The Double Clue		The Double Clue
The King of Clubs	Hit on head	The King of Clubs
The Lemesurier Inheritance	Formic acid	The Lemesurier Inheritance
The Lost Mine		The Lost Mine
The Plymouth Express	Stabbed	The Plymouth Express
The Chocolate Box	Trinitrine★★★	The Chocolate Box
The Submarine Plans		The Submarine Plans
The Third Floor Flat	Shot	The Third Floor Flat
Double Sin		Double Sin
The Market Basing Mystery		The Market Basing Mystery
Wasps' Nest	Cyanide★★	Wasps' Nest
The Veiled Lady		The Veiled Lady
Problem at Sea	Heart attack Stabbed	Problem at Sea
How Does Your Garden Grow?	Strychnine	How Does Your Garden Grow?
Curtain: Poirot's Last Case	Arsenic Morphine Shot Cyanide Hit on head Physostigmine Amyl nitrite★★★	*Curtain: Poirot's Last Case*
Sleeping Murder	Strangled Sleeping tablets	*Sleeping Murder*

UK title	Methods of murder	US title
Miss Marple's Final Cases		
Sanctuary	Shot	
Strange Jest		
Tape–Measure Murder	Strangled	
The Case of the Caretaker	Strophanthin	
The Case of the Perfect Maid		
Miss Marple Tells a Story	Stabbed	
The Dressmaker's Doll		
In a Glass Darkly`		
Problem at Pollensa Bay		
Problem at Pollensa Bay		
The Second Gong	Shot	
Yellow Iris	Cyanide	
The Harlequin Tea Set	Poisoned★★	
The Regatta Mystery		
The Love Detectives	Hit on head	
Next to a Dog		
Magnolia Blossom		
		The Harlequin Tea Set
	Fell off cliff	The Edge
		The Actress
	Shot★	While the Light Lasts
		The House of Dreams
		The Lonely God
		Manx Gold
		Within a Wall
	Stabbed	The Mystery of the Spanish Chest
	Poisoned★★	The Harlequin Tea Set

UK title	Methods of murder	US title
While the Light Lasts		
The House of Dreams		
The Actress		
The Edge	Fell off cliff	
Christmas Adventure		
The Lonely God		
Manx Gold		
Within a Wall		
The Mystery of the	Stabbed	
Baghdad Chest		
While the Light Lasts	Shot★	

Appendix 2:
Structures of some of the chemicals in this book

Some of the structures of the poisons and other chemicals discussed in this book are quite large and don't fit conveniently into short brackets that can be included in the text. A selection of structures are displayed here, so they can be compared for similarities and differences. A complete list can be found on my website, www.harkup.co.uk

B is for Belladonna

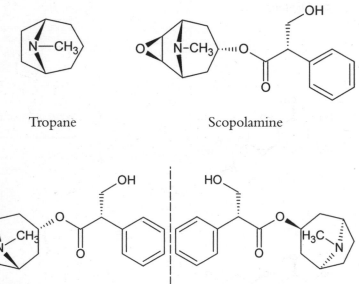

Tropane   Scopolamine

l-Hyoscamine   *d*-Hyoscamine

C is for Cyanide

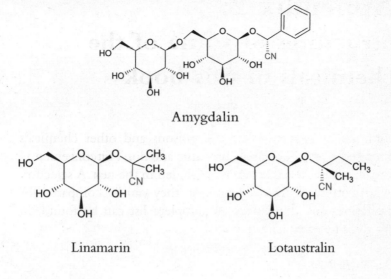

Amygdalin

Linamarin   Lotaustralin

D is for Digitalis

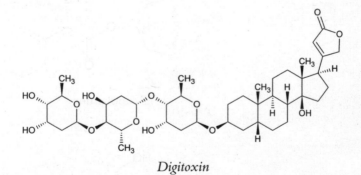

Digitoxin

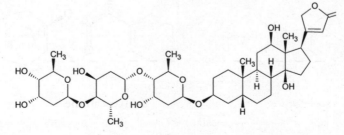

Digoxin

H is for Hemlock

Piperidine

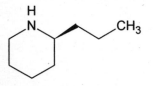

Coniine

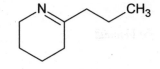

γ-coniceine

N is for Nicotine

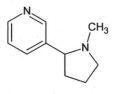

Nicotine

O is for Opium

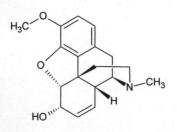

Codeine

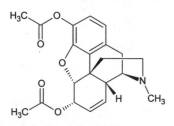

Heroin

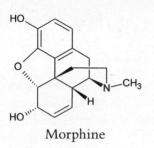

Morphine

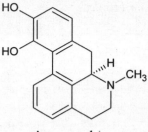

Apomorphine

V is for Veronal

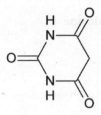

Barbituric acid

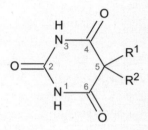

A barbiturate unit

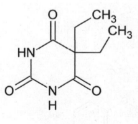

Veronal

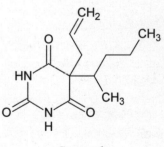

Seconal

Selected Bibliography

Below is a selection of interesting books related to the topics I've discussed in this book. A full list of citations (including many more detailed references from the academic literature) can be found on my website, www.harkup.co.uk. The website also includes the chemical structures of all of the molecules discussed in this book. Have a look.

Bereanu, V. & Todorov, K. 1994. *The Umbrella Murder*. Pendragon Press, Cambridge.

Blum, D. 2011. *The Poisoner's Handbook: Murder and the Birth of Forensic Medicine in Jazz Age New York*. Penguin, New York.

Christie, A. 1977. *An Autobiography*. William Collins Sons & Co. Ltd., London.

Cook, C. 2013. *The Agatha Christie Miscellany*. The History Press, Gloucestershire.

Curran, J. 2010. *Agatha Christie's Secret Notebooks*. HarperCollins, London.

Curran, J. 2011. *Agatha Christie's Murder in the Making*. HarperCollins, London.

Duffus, J. H. & Worth, H. G. J. 1996. *Fundamental Toxicology for Chemists*. The Royal Society of Chemistry, Cambridge.

Emsley, J. 2001. *The Shocking History of Phosphorus*. Pan Books, London.

Emsley, J. 2005. *The Elements of Murder*. Oxford University Press, Oxford.

Emsley, J. 2008. *Molecules of Murder: Criminal Molecules and Classic Cases*. Royal Society of Chemistry, Cambridge.

Farrell, M. 1994. *Poisons and Poisoners: An Encyclopaedia of Homicidal Poisonings*. Bantam Books, London.

Gerald, M. C. 1993. *The Poisonous Pen of Agatha Christie*. University of Texas Press, Austin.

Glaister, J. 1954. *The Power of Poison*. Christopher Johnson, London.

Hodge, J. H. (ed.). 1955. *Famous Trials 5*. Penguin Books, London.

Holden, A. 1995. *The St Albans Poisoner*. Corgi Books, London.

Holgate, M. 2010. *Agatha Christie's True Crime Inspirations*. The History Press, Stroud.

Klaassen, C. D. (ed.). 2013. *Casarett & Doull's Toxicology: The Basic Science of Poisons*. McGraw-Hill Education, New York.

Levy, J. 2011. *Poison: A Social History*. The History Press, Stroud.

MacEwan, P. 1912. *The Art of Dispensing: A Treatise on the Methods and Processes Involved in Compounding Medical Prescriptions*. Spottiswoode and Co Ltd., London, Colchester and Eton.

Macinnis, P. 2011. *Poisons: From Hemlock to Botox and the Killer Bean of Calabar*. Arcade Publishing, New York.

McDermid, V. 2015. *Forensics: The Anatomy of Crime*. Profile Books, London.

McLaughlin, T. 1980. *The Coward's Weapon*. Robert Hale Ltd, London.

Paul, P. 1990. *Murder Under the Microscope*. Futura Publications, London.

Rowland, J. 1960. *Poisoner in the Dock*. Arco Publications, London.

Smyth, F. 1982. *Cause of Death: A History of Forensic Science*. Pan Books Ltd, London.

Stone, T. & Darlington, G. 2000. *Pills, Potions and Poisons*. Oxford University Press, Oxford.

Thompson, C. J. S. 1935. *Poisons and Poisoners*. Barnes & Noble, New York.

Thorwald, J. 1969. *Proof of Poison*. Pan Books Ltd, London

Trestrail, J. H. 2000. *Criminal Poisoning: Investigation Guide for Law Enforcement, Toxicologists, Forensic Scientists, and Attorneys*. Humana Press Inc., New Jersey.

Waring, R. H., Steventon, G. B. & Mitchell, S. C. (eds). 2002. *Molecules of Death*. Imperial College Press, London.

Wharton, J. C. 2010. *The Arsenic Century: How Victorian Britain was Poisoned at Home, Work, and Play*. Oxford University Press, Oxford.

White, P. (ed.). 2003. *Crime Scene to Court: The Essentials of Forensic Science*. The Royal Society of Chemistry, Cambridge.

Acknowledgements

First of all, thanks to Jim Martin for giving me the opportunity to write this book. Thank you to Neil Stevens for the fabulous cover and illustrations. Julia Percival not only produced the fantastic diagrams for this book but also read and gave feedback on some of the early chapters, a contribution truly above and beyond the call of duty; thank you.

The staff at the British Library have been brilliant, particularly those of the science reading room. They have done their utmost to answer all my often obscure or ridiculous questions with enthusiasm and incredible tolerance. I am enormously grateful to Justin Brower for his excellent feedback on aspects of forensic toxicology and availability of poisons in the United States, and for his help in finding references.

More thanks are owed to my parents, Margaret and Mick, than can be properly expressed here. They have supported me emotionally and grammatically throughout the writing of this book. They have read every word, many of them several times over, and not complained once. Huge thanks and gratitude to both of them.

Many people have generously taken the time to read and give feedback on what I have written: Heather and Peter Back, Martin Bellwood, Matthew and Samuel Casey, David and Sharon Harkup, Helen Johnston, Angi Long, Sally Anne Lowe, Matthew May, Alan Packwood, Ashley Pearson, Áine Ryan, Steve Schneider, Helen Skinner, Richard and Violet Stutely, and Mark Whiting. Their contributions have been invaluable. Thank you to all of them. Special thanks must also go to Bill Backhouse for endless tea and (im)moral support.

Despite much checking and rechecking of facts there are bound to be a few errors, but these are mine and mine alone. If you spot any mistakes I would love to hear from you, care of the publishers.

Index